Clean Energy from Waste and Coal

ACS SYMPOSIUM SERIES **515**

Clean Energy from Waste and Coal

M. Rashid Khan, EDITOR

Texaco, Inc.

Developed from a symposium sponsored
by the Division of Fuel Chemistry
at the 202nd National Meeting
of the American Chemical Society,
New York, New York,
August 25–30, 1991

American Chemical Society, Washington, DC 1992

Sep/ae
chem

Library of Congress Cataloging-in-Publication Data

Clean energy from waste and coal /   M. Rashid Khan, editor.

p.   cm.—(ACS symposium series, ISSN 0097–6156; 515)

"Developed from a symposium sponsored by the Division of Fuel Chemistry at the 202nd National Meeting of the American Chemical Society, New York, New York, August 25–30, 1991."

Includes bibliographical references and indexes.

ISBN 0–8412–2514–1

1. Waste   products   as   fuel—Congresses.   2. Coal—Congresses.

I. Khan, M. Rashid. II. Series.

TP360.C55   1992
333.79'38—dc20   92–38386
                                             CIP

The paper used in this publication meets the minimum requirements of American National Standard for Information Sciences—Permanence of Paper for Printed Library Materials, ANSI Z39.48–1984. ∞

PRINTED IN THE UNITED STATES OF AMERICA

Foreword

THE ACS SYMPOSIUM SERIES was first published in 1974 to provide a mechanism for publishing symposia quickly in book form. The purpose of this series is to publish comprehensive books developed from symposia, which are usually "snapshots in time" of the current research being done on a topic, plus some review material on the topic. For this reason, it is necessary that the papers be published as quickly as possible.

Before a symposium-based book is put under contract, the proposed table of contents is reviewed for appropriateness to the topic and for comprehensiveness of the collection. Some papers are excluded at this point, and others are added to round out the scope of the volume. In addition, a draft of each paper is peer-reviewed prior to final acceptance or rejection. This anonymous review process is supervised by the organizer(s) of the symposium, who become the editor(s) of the book. The authors then revise their papers according to the recommendations of both the reviewers and the editors, prepare camera-ready copy, and submit the final papers to the editors, who check that all necessary revisions have been made.

As a rule, only original research papers and original review papers are included in the volumes. Verbatim reproductions of previously published papers are not accepted.

M. Joan Comstock
Series Editor

Contents

Preface

THE USE OR DISPOSAL OF OUR ENORMOUS WASTE RESOURCES in an efficient and environmentally acceptable way constitutes one of the major challenges of our time. Currently, nearly half the petroleum consumed in the United States annually is imported. The billions of dollars the U.S. spends for imported oil underlines the magnitude of this country's energy needs and rate of consumption. Both domestic consumption and dependence on imported energy are expected to grow. Therefore, conversion of our wastes into clean energy is a significant—and urgent—technical challenge for our scientists.

More than 160 million tons of garbage are produced in the United States every year, about two-thirds of a ton per person. In our culture of planned obsolescence, even durable goods find their way to the junkyard. It is often more convenient and economical to throw away clocks and cabinets than to repair them, as is generally done in other nations. Use of solid wastes, agricultural residues, and trees through thermal conversion processes have been practiced to a varying extent in many parts of the world. Energy shortages and environmental issues during the past decades, however, have introduced new perspectives in developing energy resources from these waste materials.

We need to develop environmentally acceptable and economical waste-based fuel forms by appropriate pre- and posttreatment. The objective of this book is to identify problems and opportunities in deriving clean energy from waste. The following wastes are considered: municipal solid waste, sludge, biomass, plastics, and tires. The chapters can be divided into two broad categories: (a) fundamental and applied aspects of waste to energy conversion and (b) the characterization, use, and disposal of byproduct ash. In many chapters, co-utilization of coal and waste have been considered. No book of this sort can hope to be comprehensive; we have tried to present an interdisciplinary treatment of some of the major topics to stimulate scientific collaboration.

This book would not have been possible without the dedicated work of the authors, reviewers, and the ACS Books Department staff. The efforts of Rhonda Bitterli, Cheryl Shanks, and Bruce Hawkins of the ACS are greatly appreciated.

M. RASHID KHAN
Texaco, Inc.
Beacon, NY 12508

October 1, 1992

Chapter 1

Clean Energy from Waste

Introduction

M. Rashid Khan[1] and Kenneth E. Daugherty[2]

[1]Research and Development Department, Texaco, Inc.,
P.O. Box 509, Beacon, NY 12508
[2]Chemistry Department, University of North Texas, Denton, TX 76203

It has been stated that the North Americans are short on energy but long on wastes. On a per capita basis, the United States is the largest consumer of energy but the greatest producer of waste in the world. The US cities dispose of about half a million tons of municipal solid waste (MSW) daily (about 95% of which is land-filled), while importing over 15 million barrels of crude daily. Over 80% of the cities are expected to run out of landfills by the end of this decade. The enormity of the problem has been depicted by the story of a garbage-laden barge from NY crossing international waterways in desperate search of a dump-site.

The US uses about 5% of its MSW to generate electricity, while the other industrial nations have been doing so at a much larger scale for a longer time. For example, Japan uses 26% of its waste in 361 plants. The use by other nations are as follows: Germany 35%; Sweden 51%; Switzerland over 75%. The slow growth in the waste utilization facilities in the US is primarily a result of concern over air pollution generated from these facilities.

In addition to MSW problems, to the treatment plants in the US, the total daily flow of wastewater is expected to increase from a current 28 billion gallons to about 33 billion gallons by the year 2000. Much of the sewage, present in the wastewater, and industrial sludges had been traditionally landfilled or ocean dumped.

Over a decade ago, the US embraced several waste-to-energy facilities, but plant failures and operational difficulties were the norm. Years later, the technology is now more mature but the problem still exists in terms of public perception and governmental support. There are barriers on both scientific and social/institutional aspects. For example, in response to mounting pressures on municipalities to meet increasingly stringent

environmental regulations by installing expensive control
equipment, many simply decided to retire their facilities
to "moth-ball" status. When an organization plans a
waste-to-energy plant, the community typically responds
by expressing the "not-in-my-backyard" syndrome.
There are serious concerns over the potential release
of some products of incomplete combustion (products such
as benzopyrene, dioxin, and benzofurans) and emission of
volatile metals from incinerators. Numerous studies have
been initiated by the US DOE, EPA and various industries
for identifying ways to mitigate these problems. Energy
recovery   from   waste   has   received   broad   technical
acceptability   as   new   and   advanced   technologies
demonstrate their capability. For example, in contrast
to direct combustion, advanced gasification technologies
can be used to produce a clean synthesis gas from waste.
The synthesis gas can be combusted in advanced gas
turbines to generate electricity without contributing to
air pollution. Coal represents 90% of the U.S. proven
reserves of fossil fuels. In many cases, both waste and
coal are used for energy generation. Co-utilization of
waste and coal offers many technical and economic
advantages not achieved by processing waste alone. Thus,
in several chapters of this book, the use of coal has been
considered along with waste.
The   objective   of   this   book,   contributed   by
internationally recognized experts, is to present the
state-of-the-art data related to waste-to-clean energy
processes. The major wastes of interest are: sewage and
industrial sludge, municipal solid waste, tires, biomass,
plastics and polymeric wastes. The reader will recognize
that several other important wastes (refinery waste, black
liquor, etc.), may well serve as excellent sources of
energy, but are not considered here.

Sewage and Industrial Sludge

Dumping of sludge in the Boston Harbor became a hot debate
issue during the 1988 presidential election, although a
relatively small portion of sludge produced is ocean
dumped (Table I). Major cities in the Eastern US (New
York City, Boston, and municipalities located in major
counties such as Westchester and Suffex) had been dumping
sewage sludge in the ocean. Treatment plants in the US
produce about 7 million tons of dry sewage sludge daily.
About 20% of the sludge produced is incinerated and about
60% of it is landfilled, while only a small portion of the
overall production is ocean dumped. The EPA, however, has
imposed a stiff penalty for ocean dumping of sludge.
After January, 1992, the penalty for ocean dumping was
increased to $600 per dry ton of sludge in the Eastern US.
The alternatives of ocean dumping, namely landfilling or
incineration,   are   meeting   strong   opposition   for
environmental reasons.

Table I. Disposal and End Use of Sludge

Method	(dry ton/y) Amount	Percent
Landfilling (dumping)	3,094	40.8
Incineration	1,541	20.3
Land application (usage)	1,424	18.8
Ocean disposal	498	6.6
Distribution/marketing	455	6.0
Other	541	7.7

SOURCE: Adapted from ref. 1.

Before sludge can be burned or gasified autogeneously (without auxiliary fuels) some level of water removal is needed. The major technical steps involved for the use of sewage sludge as a source of energy are the following: dewatering, conditioning, and processing to recover energy by combustion or gasification. For example, processes using the multiple hearth furnace, fluid-bed combustor, or one of the many gasification technologies can be readily used to extract energy from sludge.

Primary sewage sludge is generally thickened by gravity, but the sludge can also be mechanically dewatered to produce a relatively dry cake. Secondary sludge is generated by the biological treatment processes. The typical heat content of the primary sludge is about 7000 Btu/lb, which is greater than that of the secondary sludge which contains a heating value of about 4000 Btu/lb. The solids content of the primary sludge is low, generally lower than 10 percent. The types of water that can be associated with a particle of sewage sludge can be classified as follows: water bound to oxygenated functional groups and/or cations, capillary or micropore bound water, monolayer water, and bulk or free water. Polymeric additives are generally added to agglomerate the sludge particles and thereby water removal is facilitated. About 3 to 8 lbs of polymer is used per ton of dissolved solids. Polymer usage is not only expensive, but added polymers may also adversely affect the slurrying and pumping characteristics of the dewatered sludge. The major technology currently in use is incineration. Land usage of sludge, however, may require some conditioning of sludge to lessen the biological activity. Pyrolysis based processes apply relatively high temperatures to produce a liquid fuel from sewage sludge. However, no such process has ever been commercialized. Because of concerns related to environmental issues, incineration has become an unpopular option to dispose of sludge. In particular, the fate of heavy metals present in sludge during incineration is not well known. There are concerns regarding the release of some heavy metals into the atmosphere. Wet air oxidation processes serve to dewater the sludge but do not address the issue of ultimate disposal of sludge or the metals present in it. During gasification, however, the organics present in sludge are converted to useful synthesis gas (CO & H_2) while the minerals present in the feed are converted into rigid non-leachable slag potentially useful in the materials industry.

The major technical issues associated in processes aimed at extracting energy from sewage and industrial sludges are the following: (a) low-cost dewatering operation; (b) improvement of the rheological and pumpability of concentration sludge (containing >20% solids); (c) nature, type and fate of inorganics present in sludge during advanced processing; (d) improvements in emission and odor control during sludge processing. Some interesting papers on sewage sludge can be found in references 1-5. Details of the issues are beyond the scope of this chapter.

Municipal Solid Waste

Municipal Solid Waste (MSW) is one of the least used by-product resources in the United States (6,7). Until recently the common practice for disposal of, MSW was dumping. The dumping sites were usually either the ocean or an open pit (8,9). The disposal of refuse is an increasing concern of municipalities and state governments throughout the United States. In the year 1990, it was estimated that 160-200 million tons of MSW was disposed from the residential, commercial, and institutional sectors and is increasing yearly (9). Each ton of municipal solid waste is equivalent in energy content to more than a barrel of oil (10).

Cities are running out of space for landfills. One of the attractive solutions to landfills is incineration. In the early 1970s, environmental concern began to rise causing citizens to become increasingly cautious of residing near landfill sites. Due to air pollution, the garbage or MSW was no longer burned. At that time there were many landfills, and new landfill sites were available for disposal of garbage. Those landfills are either full or becoming full, and new landfills are expensive and difficult to site.

Americans dispose of 80-90 percent of their MSW into landfills filling them very quickly. The landfills in America have been reduced from 10,000 in 1980 to 6,500 in 1988, to less than 5,000 in 1992 (11,12,13). Not only is the air polluted because of MSW, but the ground water is polluted as well when the garbage decomposes. Changes in the weather and rainfall are major factors contributing to garbage decomposition. It has been estimated that water polluted today will be affected for hundreds of years (14).

In 1989, the Environmental Protection Agency (EPA) proposed regulations for stricter control of new and preexisting landfills. These measures, which went into effect in 1991, will help in solving the problem, but are expensive. It is estimated that it will cost over 800 million dollars per year to implement these methods nationwide. The regulations and controls include monitoring ground water for contamination, allowing for the controlled escape of methane which forms as the garbage decomposes, and permanently sealing landfills after they are filled (15).

Sources of Municipal Solid Waste. Municipal refuse is a heterogeneous mixture of organic and inorganic wastes discarded by homes, schools, hospitals, and a variety of other sources in the community. The major contributors to solid waste are (16):

a)   Domestic: single and multiple dwellings
b)   Commercial: offices and retail stores
c)   Entertainment centers: restaurants, hotels and motels, and service stations
d)   Institutional: schools, hospitals and municipal buildings
e)   Municipal services: demolition and construction, street and alley cleaning, landscaping, catch basin cleaning, parks and beaches, and waste treatment residues

Content of Municipal Solid Waste. Municipal solid waste is an aggregate mixture of waste materials that can be classified as an organic fraction, an inorganic fraction, and moisture. The organic fraction, which makes up to 30% of the waste, is primarily cellulose (wood fibers). It is considered a major source for energy recovery. The inorganic fraction is noncombustible. It can be either recyclable or after combustion constitutes the ash residue. Table II shows the summary of the chemical characterizations of MSW (17).

Another aspect or objective of many of the recovery processes of MSW is to utilize its thermal energy. The heat content of MSW is important. The heat content of the as-received refuse can reach 3,500 to 5,500 Btu/pound (16). A reduction of the moisture or inert contents will increase the heat content. Decreased quantities of plastics will also decrease the heat content of MSW (18).

There are many solutions that have been proposed for the problem of growing landfills and the increase in MSW. These solutions include mass burning, burying, recycling, and use as an energy source using Refuse Derived Fuel (RDF).

The analyses shown in Table II were developed by one of the authors during experimental work on the Denton, Texas landfill during 1980-81 (17). The composition was based on repeated sampling and testing of the Denton landfill.

Table II.   Municipal Solid Waste Composition (moisture
            free) in Denton, Texas - 1981

a.   Combustible
            Paper   52%
            Plastic   14%
            Wood   5%
            Garden Waste   4%
            Food Waste   3%
            Rubber   1%
            Leather   1%

b.   Non-Combustible
            Glass/Ceramic/
            Stone   9%
            Ferrous   6%
            Aluminum   2%
            Industrial/
            Commercial   2%
            Residual
            Dirt   1%

SOURCE: Reprinted from ref. 17.

Table III presents results of the manufacturing operations
of the Eden Prairie Recycling plant in Eden Prairie,
Minnesota (Division of Green Isle Environmental Services,
Inc.).   The data presented are recycling data for the
months of June, July and August of 1992 (19).

Conversion or Oxidation of Municipal Solid Waste.
Burning MSW not only reduces the volume of garbage by 80%
but may also provide usable energy.   MSW can be used in
three different ways:

1.   Direct combustion, mass burn: the MSW is fed into
        the furnace through a moving grate where the
        temperature reaches 2400°F.   The problems with the
        mass burn incinerators are the cost of the
        incineration facility and the emissions.

2.   Conversion of MSW into liquid or gaseous fuel by
        means of gasification, pyrolysis, biodegradation,
        or hydrogenation. The liquid or gaseous   fuel
        produced can then be easily cofired with coal or
        oil.

3.   Burning of the combustible portion of MSW, Refuse
        Derived Fuel   (RDF),   after separating   the
        incombustible portion.

   The incombustible portion of MSW and the ash are
discarded in landfills which create a new problem, namely
pollution.   Ash contains some organic constituents and
some trace elements at different levels.   Ash is
considered hazardous if the levels of toxic constituents
are high.

Table III. Eden Prairie Recycling, Inc. (19)				
		1992 Tons		Total	Total
	June	July	August	Tons	%

Recyclables

	June	July	August	Total Tons	Total %
Aluminum	16.40	14.11	10.73	41.24	0.15
Metal	0.00	0.00	0.00	0.00	0.00
Plastic	3.21	3.12	5.54	11.87	0.04
Corrugated	170.81	191.47	121.24	483.52	1.71
Ferrous Mtl	114.87	95.51	102.14	312.52	1.11
Scrap Metal	107.18	91.37	79.56	278.11	0.99
TOTALS	412.47	395.58	319.21	1127.26	4.00

Fuel

	June	July	August	Total Tons	Total %
RDF	2982.16	3376.23	2563.49	8921.88	31.62

Composted/Incinerated/Other

	June	July	August	Total Tons	Total %
	4091.09	3170.64	2286.46	9548.19	33.84

Landfilled

	June	July	August	Total Tons	Total %
Heavies	1714.97	3246.49	1992.34	6953.80	24.65
Rejects	496.55	609.67	557.72	1663.94	5.90
TOTALS	2211.52	3856.16	2550.06	8617.74	30.55
TOTAL TONS	9697.24	10798.61	7719.22	28215.07	100.01

Table IV shows that the use of MSW as a source of renewable energy is expected to grow at a rate higher than 8% between 1989 and 2010 (20).

Landfilling MSW. Sanitary landfills were increased in 1976, when the Resource Conservation and Recovery Act (RCRA) gave the EPA the authority to close open landfills and upgrade the quality of sanitary landfills. Sanitary landfills are typically huge depressions lined with clay to minimize leakage of pollutants into the groundwater. Heavy equipment is used to spread the MSW out and compress it every day. After the landfill has been packed to capacity, a layer of dirt and/or plastic is used to cover the day's haul.

Table IV. Renewable Energy
(Quadrillion Btu per Year)

Electricity and Non-Electric	1989	2010	Annual Growth 1989-2010 %
Electricity Capability (gigawatts)			
Convent. Hydropower	75.48	78.46	0.2
Geothermal	2.47	10.65	7.2
MSW	1.98	10.81	8.4
Biomass/Other Waste	5.31	8.88	2.5
Solar Thermal	.33	1.78	8.4
Solar Photovoltaic	.00	.01	2.1
Wind	1.93	5.30	4.9
Total	87.51	115.90	1.3
Generation (billion Kilowatthours)			
Convt. Hydropower	276.90	314.80	.6
Geothermal	15.05	78.52	8.2
MSW	13.31	74.22	8.5
Biomass/Other Waste	29.54	49.55	2.5
Solar Thermal	.69	5.06	10.0
Solar Photovoltaic	.00	.00	2.2
Wind	3.38	12.97	6.6
Total	338.90	535.10	2.2
Consumption/Displacement			
Convent. Hydropower	2.88	3.27	.6
Geothermal	.16	.82	8.2
MSW	.20	1.27	9.2
Biomass/Other Waste	.20	.33	2.5
Solar Thermal	.01	.05	10.0
Solar Photovoltaic	.00	.00	2.3
Wind	.04	.13	6.6
Total	3.48	5.88	2.5
Non-Electric Renewable Energy **Residential, Commercial & Industrial**			
Hydropower	.00	.00	–
Geothermal	.00	.39	–
Biofuels	2.63	4.54	2.6
Solar Thermal	.05	.54	11.6
Solar Photovoltaic	.00	.00	–
Wind	.00	.00	–
Transportation			
Ethanol	.07	.14	3.5
Total	2.75	5.62	3.5
Total Renewable Energy	6.23	11.50	3.0

SOURCE: Reprinted from ref. 20.

Sanitary landfill operators follow strict guidelines. They control and monitor methane gas generation, surface water runoff, and groundwater contamination by the landfill.

Refuse Derived Fuel

Refuse Derived Fuel (RDF), shredded MSW with most glass and metals removed, is an attractive solution since it also addresses another problem affecting the United States: depleting energy reserves. One ton of RDF has the energy equivalent of one barrel of oil (21). RDF has a 7,000 to 8,000 Btu/pound heat content. The powdered RDF, embrittled and pulverized refuse, will even have a higher heat content of over 8,500 Btu/pound (16).

RDF Technology. The starting material of RDF is MSW. The exact composition of MSW varies according to the area, the time of the year it was collected, and the make-up of that particular community. RDF refers to the heterogenous mixture of the combustible portion of MSW (22).

The concept of RDF has existed since the early 1970s (23). There are seven forms of RDF that have been defined as described in Table V (22). RDF is commonly used in two forms, fluff (RDF-1) and densified RDF (RDF-5; dRDF).

There are several problems with using RDF-1 that make it less attractive, such as: it is hard to handle, it is usually burned in suspension: and often causes problems in ash handling since much of it remains unburned (14). On the other hand, the main benefits of using RDF rather than raw refuse are:

* RDF when properly processed, can be stored for an extended period of time

* RDF technology allows for the recovery of saleable material

* RDF can be combusted in a wide range of existing boilers, fluidized bed combustors, gasifiers, and cement and brick kilns

* RDF can also be used as a feedstock for anaerobic digesters to produce methane gas

* RDF can easily be transported from one location to another

* RDF can be burned on a supplemental basis with other fuel, such as coal or wood

Table V.   Types of Refuse Derived Fuel

RDF-1   waste used as fuel in as-discarded form

RDF-2   waste processed to coarse particle size with or without ferrous metal separation

RDF-3   shredded fuel derived from MSW that has been processed to remove metals, glass and other inorganic materials (95 wt% passes 50-mm square mesh)

RDF-4   combustible waste processed into powder form (95 wt% passes 10 mesh)

RDF-5   combustible waste densified (compressed) into a form of pellets, slugs, cubits or briquettes (dRDF)

RDF-6   combustible waste processed into liquid fuel

RDF-7   combustible waste processed into gaseous fuel

SOURCE: Adapted from ref. 22.

* RDF is more homogeneous, yielding low variability in fuel characteristics, thereby making combustion control easier to implement. It also burns more evenly at a higher sustained temperature.

* RDF has a lower percentage of unburnable residuals such as metals and glass, and this has a higher heat content per unit weight than does unprocessed solid waste.

* RDF when burned in a dedicated boiler has a greater thermal efficiency (8-10 percent greater).

* RDF can have a beneficial effect on air emissions and ash residue, compared to burning MSW.

In order to effectively utilize the combustible portion of MSW, known as RDF, it is necessary to densify the RDF in order to transport it economically and easily. It is then called Densified Refuse Derived Fuel (dRDF). This densification step can increase the density of RDF from 2 to 3 pounds per cubic foot to 40 or more pounds per cubic foot (10). If dRDF is going to be stored for a period of time longer than several days, a binder must also be added. Calcium hydroxide ($Ca(OH)_2$), which has been proven to be the best binder, is often added to RDF before densification. The binder delays biological and chemical degradation for years and reduces SO_x, NO_x and other emissions by cofiring with coal (12,23).

Plastics

Plastics, unrecyclable and recyclable, make up 9% of MSW by weight and approximately 19% by volume. In 1987, over 57 million pounds of plastics were sold in the US. Although plastics have been recycled for over a decade, only a little over 1% of the plastics produced are recycled. Even with a large increase in recycling, there is considerable room for research and development to find ways to derive clean energy from plastics.

Governments have taken or are considering taking actions that affect plastic recycling. A significant one involves the banning of certain plastics outright. One target of such laws is polystyrene, the foamed, insulating plastic that is most often used for coffee cups and fast food packaging. Some areas have banned polystyrene packing; others have banned polystyrene foam made with chlorofluorocarbons as a blowing agent. The US Congress has been considering measures that would encourage recycling by placing a tax on the use of virgin materials. If industries could successfully demonstrate the usage of these materials as a feedstock for making useful chemicals or energy, then the urgency to ban making them would be reduced. Once the technical issues of feed collection and preparation were addressed, these wastes could be used as a source of clean energy.

Virtually each type of plastic has unique chemical and physical properties. These qualities effect the recyclability of the plastics and the use of recycled plastics. For example, polyurethane mixed into a batch of polyethylene terephthalate (PET) can greatly reduce the strength of the PET and render it unsuitable for many applications. That means, the recyclers must separate the different types of plastics before doing anything with them. This also shows that there are limits on the recyclability of plastics.

With many plastics being unrecyclable, the processing required for recycling often being limiting, and the fact that plastics have a high energy content (18,000 Btu/lb.), there exists an opportunity for plastic waste as a fuel. The current options for plastics, in addition to recycling, are combustion (incineration) and landfilling, both of which are unpopular. Successful use of plastics as a feedstock for a process such as gasification would help to reduce landfilling and incineration and would provide a clean source of energy.

Other Wastes

There are numerous waste materials which can be considered as potential clean energy sources, in addition to the sludge, MSW and plastics considered above. Tires, for example, are being discarded in the United States at the rate of about one tire per person per year. Tire rubber has an energy content of about 18,000 Btu per pound but has some environmental concerns from a combustion perspective.

Concluding Remarks

Each waste material is a problem requiring extensive study and assessment of sound technology. Nor are these problems mutually exclusive: as (for example) tires and biomass are commonly encountered in MSW. However, each major waste area has the potential of becoming a waste-to-clean energy process. Through the dedication and efforts of recognized experts in this book and elsewhere, technologies will be developed to produce clean energy from wastes.

Literature Cited

1.   "Economic Analysis of Sewage Sludge Disposal Alternatives", Proc. of National Conference on Municipal Sewage Sludge Treatment Plant Management, May, 1987. (This table was revised)
2.   Chow, V.T., Eliassen, R., and Linsley, R.K., "Wastewater Engineering: Treatment Disposal Reuse", McGraw-Hill, 1979.
3.   "National Sewage Sludge Survey: Availability of Information and Data, and Anticipated Impact on Proposed Regulations; Proposed Rule", Federal Register, Vol. 55, No. 218, 40 CFR Part 503, November 9, 1990.
4.   Zang, R.B. and Khan, M.R., "Gasification of Sewage Sludges", presented to New York Water Pollution Control Association Annual Meeting, January, 1991.
5.   Makansi, J., "Power from Sludge", Power, February, 1984.
6.   Attili, B., "Particle Size Distribution and Qualitative/Quantitative Analysis of Trace Metals in the Combustion of Gas and Fly Ash of Coal/Refuse Derived Fuel", Ph.D. Dissertation, University of North Texas, Denton, Texas, December, 1991.
7.   Hasselriis, Floyd, Refuse Derived Fuel Processing, Butterworths Publishers, Boston, 1983.
8.   Hill, R., Daugherty, K.E., Zhao, B. and Brooks, C., "Cofiring of Refuse Derived Fuel in a Cement Kiln", International Symposium on Cement Industry Solutions to Waste Management, Calgary, Canada, October, 1992.

9.   Ohlsson, O., and Daugherty, K.E., "Results of Emissions in Full Scale Co-Combustion Test of Binder Enhanced d-RDF Pellets and High Sulfur Coal", presented at Air and Waste Management Association Forum 90 in Pittsburgh, Pennsylvania, 1990.

10.   Daugherty, K.E., Refuse Derived Fuel, Monthly reports to the US Department of Energy, University of North Texas, Denton, Texas, 1984-1986.

11.   Rice, F., Fortune, April 11, 1988, 177, pp. 96-100.

12.   Hill, R. and Daugherty K.E., "Binder Enhanced Refuse Derived Fuel - Possible Emission Reductions", accepted for publication by Solid Waste and Power, Kansas City, Missouri, publication slated for November/December, 1992.

13.   Reuter, Inc., Annual Report, 1991.

14.   Diegmueller, K., Insight, 1986, 12, pp. 16-17.

15.   Johnson, P., McGee, K.T., USA Today, August 26, 1988.

16.   Hecht, N., Design Principles in Resource Recovery Engineering, Ann Arbor Science, Butterworth Publishers; Boston, 1983, pp. 23-33.

17.   Daugherty, K.E., "RDF As An Alternative Source of Energy for Brick Kilns", Department of Energy, DOE/CS/24311-1, December, 1982.

18.   Gershman, Brickner, and Bratton, Inc., Small Scale Municipal Solid Waste Energy Recovery Systems, Van Nostrand Reinhold Company, New York, 1986, pp. 4-20.

19.   Personal communication between Dr. Kenneth Daugherty and Mr. John Schilling, Assistant Plant Manager of EPR, Inc. on September 11, 1992.

20.   US Department of Energy, Energy Information Administration/Annual Energy Outlook 1991.

21.   Carpenter, B., Windows, Spring 1988, pp. 8-10.

22.   Alter, H., Material Recovery from Municipal Solid Waste, Marcel Dekker, Inc., New York, 1983, pp. 181-190.

23.   Daugherty, K.E., "An Identification of Potential Binding Agents for Densified Fuel Preparation from Municipal Solid Waste, Phase 1, Final Report", Argonne National Laboratory, Argonne, Illinois, 1988.

RECEIVED October 6, 1992

Chapter 2

Efficient and Economical Energy Recovery from Waste by Cofiring with Coal

Charles R. McGowin and Evan E. Hughes

Electric Power Research Institute, 3412 Hillview Avenue,
P.O. Box 10412, Palo Alto, CA 94303

Cofiring fuels derived from municipal and nonhazardous industrial wastes with coal in industrial and utility boilers is an efficient and cost-effective method of energy recovery from wastes in many cases. Waste fuels such as scrap tires, tire-derived fuel, refuse-derived fuel, paper mill sludge, sewage sludge, sawdust, wood, and industrial waste can be cofired with coal in many stoker, pulverized coal, cyclone, and fluidized bed boilers with only minimal modifications and with minimal impacts on environmental emissions and plant safety. Waste cofiring with coal usually exhibits a higher waste-to-energy conversion efficiency than 100 percent waste firing in dedicated waste-to-energy plants, because coal-fired plants typically operate at higher steam pressures and temperatures and therefore higher steam-cycle and thermal efficiencies than dedicated plants. In addition, waste cofiring generally requires a much lower incremental capital investment than waste firing in a dedicated waste-to-energy facility. Both factors can contribute to a lower breakeven waste disposal cost or tipping fee for waste fuel cofiring with coal than for dedicated plants. This economic advantage should be highest for low-volume, low heating-value fuels, such as municipal solid waste and sewage sludge, and lowest for high-volume, higher quality fuels, such as scrap tires.

In response to the environmental crisis in the U.S. created by the growing volume of municipal and industrial wastes and declining availability of landfill disposal sites, many urban communities are developing integrated waste management plans to both reduce the volume of the wastes sent to landfill and recover valuable raw materials and energy as steam and/or electricity. Most integrated waste management plans involve a combination of recycling, composting, waste-to-energy technology, and landfilling of residues, applied in sequence. Waste-to-energy plants will therefore be the last stop for the large portion of the waste stream that is not recyclable.

The waste fuels burned in waste-to energy plants are derived from a variety of sources, including residential and commercial refuse, sewage sludge, automotive tires, urban demolition wastes, agricultural wastes, wood waste from forestry operations and lumber mills, paper mill sludge, and other industrial wastes. Most waste-to-energy

0097–6156/93/0515–0014$06.00/0

facilities use dedicated waste-to-energy technology designed to efficiently recover energy as steam or electricity while controlling environmental emissions. In addition, several existing fuel combustion facilities have been retrofitted to burn waste fuels either alone or in combination with coal or oil, including industrial and utility boilers and cement kilns.

This paper addresses the current and future generation of waste fuels in the U.S., waste fuel properties, alternate waste-to-energy technologies, and energy conversion efficiencies and costs of the alternate technologies. To represent the range of waste fuels available, the discussion focuses on scrap tires, refuse-derived fuel, wood waste, and sewage sludge.

Annual Generation and Properties of Waste Fuels

Of the 185 million tons of municipal wastes and 200 million waste tires generated each year, it is estimated that up to 25 percent could be recycled and reused, leaving about 75 percent for disposal in landfills and waste-to-energy facilities. The remaining 140 million tons of MSW contains the energy equivalent of 58 million tons of bituminous coal and would be sufficient to provide 11 thousand megawatts of generating capacity. The corresponding figures for waste tires are 1.9 million tons of coal and 640 MW of generating capacity. Currently, about 16 percent of the municipal solid waste stream and a small fraction of the waste tires are processed in waste-to-energy facilities, and these fractions are expected to grow significantly by the year 2000, perhaps to 40 percent. At the same time, annual waste generation can be expected to grow at about two percent per year (*1*).

Typical fuel properties are presented in Table I for tire chips, refuse-derived fuel, wood waste, and sewage sludge (*2,3*). The rubber tire chips are made by shearing waste tires to one-inch top size and removing as much of the steel belt and bead material as possible by magnetic separation, producing a fuel containing 1.2 percent sulfur, 14.8 percent ash, and 12,500-14,500 Btu/lb. The refuse-derived fuel (RDF) is produced from municipal solid waste by shredding, screening, and magnetic separation and contains 0.2 percent sulfur, 12 percent ash, 24 percent moisture, and 5,900 Btu/lb. The wood waste is a mixture of chipped forest residue, bark, and milling waste, containing 0.02 percent sulfur, 0.7 percent ash, 39 percent moisture, and 5,140 Btu/lb. The municipal sewage sludge has been dewatered to 86.2 percent moisture content in a centrifuge and contains 6.8 percent ash. Due to its high moisture content, the high heating value is only 464 Btu/lb and the low heating value is negative (-484 Btu/lb).

Table I. Coal and Waste Fuel Properties

	WV Coal	Tire Chips	RDF	Wood Chips	Sewage Sludge
% S:	0.85	1.19	0.20	0.02	0.07
% Ash:	16.04	14.76	12.0	0.74	6.8
% Moisture:	6.60	8.55	24.0	39.10	86.2
Btu/lb:	11,680	13,500	5,900	5,139	464

SOURCE: Adapted from ref. 2.

Due to their low sulfur contents, cofiring wood chips or RDF with coal would reduce emissions of sulfur dioxide (SO_2) relative to those for coal firing. However, as discussed below, the high moisture contents of wood chips and RDF lead to degradation of power plant efficiency and performance. Relative to heat content, tire chips and sewage sludge have moderate sulfur contents, and cofiring with coal can either increase or decrease SO_2 emissions, depending on the coal sulfur content.

Waste-to-Energy Technologies

Several methods are available for energy recovery from waste fuels, including construction of dedicated waste-to-energy facilities and conversion of existing coal-fired facilities to burn the waste fuels. Most dedicated waste-to-energy facilities use mass-burn, stoker firing, or fluidized bed combustion technology to burn the waste fuels (4,5). In mass burn plants (Figure 1), waste fuels are typically burned in a refractory lined furnace or on a sloping reciprocating grate without prior size reduction or processing to remove noncombustibles. Stoker-fired and fluidized bed plants typically require at least some size reduction to avoid plugging the fuel handling and injection equipment (4). In Japan, several fluidized bed units are burning industrial and municipal wastes without significant size reduction (4,6).

Many existing facilities, including industrial boilers, cement kilns, and coal-fired power plants, can be converted to burn waste fuels either alone or in combination with other fuels such as coal. Figure 2 is a schematic of a utility boiler converted for RDF cofiring. Currently, RDF is cofired with coal at one cyclone and three pulverized coal power plants (7), cofired with wood at two fluidized bed plants (8,9), and fired alone at three converted utility stoker-fired plants (10). Whole tires are cofired with coal at one wet-bottom pulverized coal plant, and tire-derived fuel has been cofired with coal at four cyclone-fired plants and at three stoker-fired plants (11). In addition, one utility has cofired pulverized wood chips with coal (12).

It should be noted, however, that cofiring alternate fuels is not always technically feasible and that the maximum heat input fraction is often limited by practical or economic considerations. Factors that need to be considered include the boiler type (pulverized coal, cyclone-fired, stoker-fired, or fluidized bed), operating and performance limitations on coal, and required fuel specifications.

Pulverized coal boilers are designed to burn finely-ground coal in suspension. Most must be modified with a bottom dump grate to handle solid fuels such as RDF and tire and wood chips, although, as mentioned earlier, whole tires have been successfully tested in a slagging wet-bottom boiler without a dump grate (11). Cyclone-fired units burn the solid fuels in external cyclone burners, and because they remove most of the ash as a molten slag in the burners, a bottom dump grate is not required. It is unlikely that more than 20 percent of the heat input could be provided by alternate fuels in either case, due to adverse impacts on burner performance and heat release and absorption profiles in the boiler. Stoker-fired and fluidized bed combustion units would be generally more suited to cofiring alternate fuels at higher heat input fractions, subject to the operating and performance limitations discussed below.

Some utility boilers may exhibit operating and performance limitations on the net generating capacity while firing coal (13,14). These include the boiler convection pass flue gas velocity, ID fan capacity, electrostatic precipitator performance, ash handling capacity, and boiler tube slagging and fouling. In this case, cofiring low grade fuels such as RDF, wood chips, or sewage sludge would likely result in a loss of net generating capacity, even at low heat input fractions. This is because cofiring increases flue gas and bottom ash volumes and worsens any existing slagging and fouling problems. These boilers would not be suitable for cofiring alternate fuels.

The degree of fuel processing, which affects fuel particle size and ash, moisture, and heat contents, can also determine the feasibility of cofiring waste fuels. High ash, glass, and metal contents can lead to increased boiler slagging and fouling in pulverized coal boilers and plugging and jamming of moving grates in stoker-fired boilers (13).

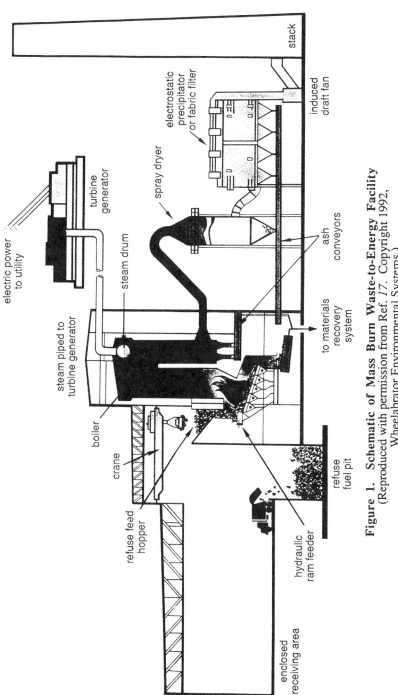

Figure 1. Schematic of Mass Burn Waste-to-Energy Facility
(Reproduced with permission from Ref. 17. Copyright 1992, Wheelabrator Environmental Systems.)

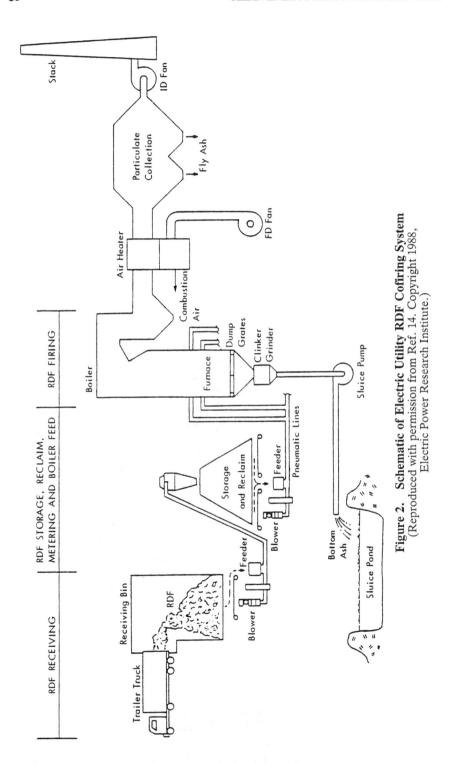

Figure 2.   Schematic of Electric Utility RDF Cofiring System
(Reproduced with permission from Ref. 14. Copyright 1988,
Electric Power Research Institute.)

Large particles can jam and plug fuel feeding and ash removal systems and cause poor combustion efficiency. Increased processing improves fuel quality and uniformity, and reduces the possibility of fuel handling and other problems.

Environmental Emissions and Plant Safety

Environmental emissions in air, water, and solid effluents from waste-to-energy facilities depend on both the chemical analysis of major components as well as the trace metal contents of the waste fuels and the emissions control equipment used by the facility. As shown in Table 1, most of the waste fuels have very low to moderate sulfur contents relative to coal and thus are not significant producers of sulfur dioxide during combustion. However, many of the wastes contain trace metals and can produce hazardous substances during combustion. For example municipal refuse is known to contain variable levels of lead, cadmium, and mercury, and chlorine derived from salt, plastics, and bleached paper can form polychlorinated dibenzo dioxins and furans under certain conditions involving poor fuel/air mixing and insufficient combustion air.

Modern dedicated waste-to-energy and coal-fired power plants are typically designed to effectively control carbon monoxide, acid gas, trace metal, and organic emissions to levels below state and federal requirement emissions limits. Wet or dry scrubbers and fabric filters or electrostatic precipitators are used to control acid gas (SO2 and HCl) and particulate/trace metal emissions. Nitrogen oxide emissions are typically controlled using low-NOx burners, combustion modification, or ammonia thermal de-NOx systems. Carbon monoxide and trace organic emissions are controlled by designing the boiler for "good combustion practice."

Waste fuel cofiring, however, can either decrease, increase, or have little impact on emissions relative to those for 100 percent coal firing. For example, sulfur dioxide emissions typically decrease when cofiring low-sulfur fuels such as RDF and wood, but may increase or decrease when cofiring moderate-sulfur fuels such as tire chips, depending on the coal sulfur content. Nitrogen oxide emissions also decrease for most waste fuels due to low nitrogen contents relative to coal and to the impact of high fuel moisture contents on flame temperature and thermal NOx formation. Particulate emissions often increase slightly for units with electrostatic precipitators due to increased flue gas volumes and fly ash resistivity for many waste fuels. It is interesting to note that no dioxin and furan emissions have been detected during stack emissions testing of utility boilers cofiring coal and refuse-derived fuel.

The primary safety concern with waste fuel cofiring is fire prevention in waste fuel, receiving, storage, and reclaim areas, particularly with highly combustible wastes such as scrap tires and dried refuse and wood.

Overall Energy Efficiency and Costs

To illustrate the overall energy efficiency and economics of waste energy recovery, simplified examples of waste fuel/coal cofired and dedicated waste-to-energy plants are presented below. These examples have been refined since they were first reported in 1991 (*15*).

Four waste fuels are considered for each system: tire-derived fuel, RDF, wood, and sewage sludge for the retrofitted coal/waste cofired plants; and whole tires, RDF, wood, and municipal solid waste (MSW) for the dedicated plants. In addition to the coal and waste fuel properties listed in Table I, major assumptions include:

- Waste fuel/coal-cofired plants:
  - 250 MW pulverized coal power plant
  - West Virginia bituminous coal
  - Steam conditions: 2400 psia/1000 deg F superheat/1000 deg F reheat
  - Wet limestone flue gas desulfurization and cold-side electrostatic precipitator
  - Retrofit fuel receiving, storage, and pneumatic transport systems and dump grate installed above ash hopper in furnace
  - 20% heat input from TDF, RDF, and wood waste; 2% heat input from sewage sludge.

- Dedicated waste-to-energy plants:
  - Same annual waste fuel consumption rates as cofired plants
  - Steam conditions: 900 psi/830 deg F.

- Reference date of capital and annual cost estimates: December 1990.

- Busbar energy cost includes fixed and variable O&M, coal, and capital charges (16.5%/yr fixed charge rate).

- Waste fuels available at zero cost.

- Zero credit for recovered byproducts such as aluminum and steel.

For waste fuel cofiring, the analysis is based in part on waste fuel cofiring performance and cost estimates reported previously (*13*). The boiler efficiencies, heat rates, thermal efficiencies, fuel consumption rates, and busbar energy costs were estimated using a revised version of the RDFCOAL RDF Cofiring Boiler Performance Model (*14*). RDFCOAL predicts the impact of RDF cofiring on boiler efficiency, net heat rate, unit derating, fuel consumption, flue gas volume and composition, and bottom and flyash production as functions of the fraction of fuel heat input provided by RDF.

For the dedicated wood-fired plant case, the performance and cost data are based on reported estimates (*16*). For the whole-tire, RDF, and MSW-fired plant cases, the data were estimated using the EPRI Waste-to-Energy Screening Guide and software (*5*). The WTE Screening Guide presents performance and cost data for alternate waste-to-energy technologies, and the software was used to prepare the estimates for the RDF/stoker fired and MSW mass burn plant cases. The estimates for the whole tire-fired dedicated plants were derived by scaling the MSW mass burn data, based on waste fuel throughput, flue gas volume, and gross MW generating capacity.

Table II and Figures 3 to 6 compare the energy recovery inefficiencies, annual waste fuel consumption rates, and incremental capital and annual costs for the cofired and dedicated waste-to-energy plants.

Boiler and Thermal Efficiencies. Energy conversion efficiency is measured by the fraction of energy in the fuel converted to steam and/or electricity, expressed as boiler efficiency and thermal efficiency, respectively.

Figure 3 compares the predicted overall boiler efficiencies as functions of waste fuel heat input fractions for utility boilers cofiring three waste fuels with coal, including tire-derived fuel, refuse-derived fuel and wood (*13*). As the waste fuel heat input fraction increases, the boiler efficiencies for RDF and wood cofiring decline significantly from 89 percent at zero percent heat input (100 percent coal firing) to 78 and 75 percent, respectively, at 100 percent heat input. This results primarily from the high moisture contents and higher excess air requirements of the waste fuels, which increase the dry

Table II. Performance and Economic Comparison of Waste Fuel-Cofired and Dedicated Power Plants (December, 1990 Dollars)

250 MW Waste Fuel/Coal-Cofired Power Plant

Assumptions: West Virginia Bituminous Coal @ $1.60/MBtu, 65% Capacity Factor

	Tire-Derived Fuel	RDF	Wood	Sewage Sludge
Waste Fuel Performance:				
Heat Input	20%	20%	20%	2%
Tons/Day	434	1018	1177	1335
Tons/Year	103,000	241,000	279,000	317,000
Thermal Efficiency	34.2%	30.5%	29.6%	8.5%
Net Btu/kWh	9,997	11,175	11,536	40,140
Incremental Total Capital Requirement:				
$/Ton/Day Waste	22,000	9500	8200	900
Breakeven Fuel Payment:				
$/MBtu	$0.03	-$0.17	-$0.13	-$4.78
$/Ton Waste	$0.74	-1.96 (RDF) -1.65 (MSW)	-1.37	-4.44
Breakeven Tipping Fee ($/Ton Waste):				
Processing Cost	$20.00	$40.00 (MSW)	$10.00	$0.00
Less Fuel Payment	(0.74)	(-1.65) (MSW)	(-1.37)	(-4.44)
B.E. Tipping Fee	$19.26	$41.65 (MSW)	$11.37	$4.44
Sensitivity Range ($2.50 to $0.50/MBtu Coal)	$-4 to 48	$34 to 51	$4 to 20	$6 to 2

Dedicated Waste Fuel-Fired Power Plant

Assumptions: Electricity Sale @ $0.05/kWh, 80-85% Capacity Factor

	Whole Tires	RDF	Wood	MSW
Waste Fuel Performance:				
Tons/Day	332	827	900	927
Tons/Year	103,000	241,000	279,000	288,000
Net Capacity (MW)	27.1	25.4	22.5	22.0
Thermal Efficiency	24.8%	21.4%	19.9%	20.2%
Net Btu/kWh	13,800	16,000	17,100	16,900
Total Capital Requirement:				
$/Ton/Day Waste	$258,000	$164,000	$54,000	$117,000
$/kW Net	$3200	$5300	$2150	$4960
Breakeven Tipping Fee ($/Ton Waste):				
Processing Cost	$205.40	$151.90	$38.80	$102.90
Less Electricity Rev.	(97.90)	(36.90)	(30.00)	(28.50)
B.E. Tipping Fee	$107.50	$115.00	$8.80	$74.40
Sensitivity Range ($0.10 to $0.03/kWh)	$9 to 147	$78 to 130	$-21 to 21	$46 to 86

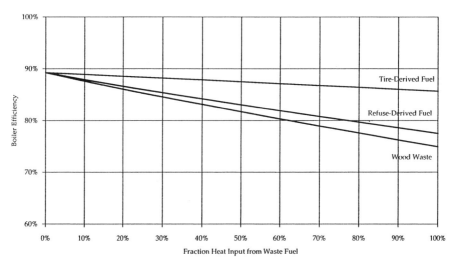

Figure 3. Boiler Efficiency for Waste Fuel/Coal Cofiring

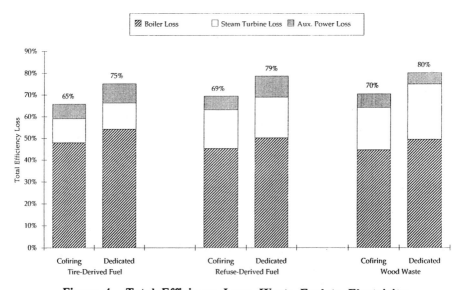

Figure 4. Total Efficiency Loss: Waste Fuel to Electricity

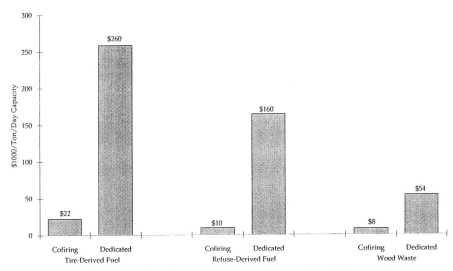

Figure 5. Total Capital Requirement for Waste Fuel Energy Recovery

Incremental Impact of Waste Processing (1990 $)

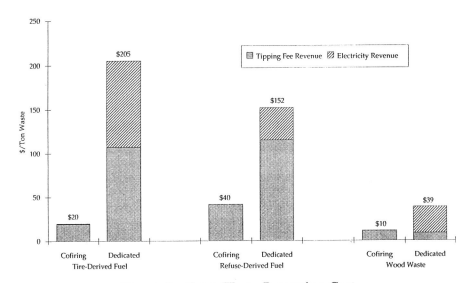

Figure 6. Total Waste Processing Cost

Coal @ $1.60/MBtu, Electricity @ $0.05/kWh, 1990 $

gas and moisture loss components of the boiler efficiency loss. For tire-derived fuel which has low moisture and high heat contents like coal, the boiler efficiency declines slowly to 86 percent at 100 percent heat input from tires. Thus, waste fuel cofiring with coal can be expected to provide higher boiler and steam conversion efficiency than 100 percent waste fuel firing, which in turn contributes to higher overall electricity conversion efficiency in most cases, as discussed further below.

The waste fuel thermal efficiency is a function of boiler efficiency, gross turbine heat rate, and auxiliary power consumption. Modern coal-fired power plants typically operate at higher steam conditions than dedicated waste-to-energy plants (e.g. 2400 psi/1000 F superheated steam with one reheat to 1000 F vs. 900 psi/830 F superheated steam) and have lower auxiliary power requirements (8% vs. 11%). As a result, the waste fuel thermal efficiency can be expected to be higher for the cofired plants than for the dedicated plants, which is confirmed by the data in Table II and Figure 4. For tires, RDF, and wood waste, the waste fuel thermal efficiencies range between 30 and 34 percent for the cofired plants and 20 and 25 percent for the dedicated plants.

Waste Fuel Consumption. Annual waste fuel consumption is proportional to capacity factor and inversely proportional to fuel heat content and thermal efficiency. The 250 MW coal plant operates at 65 percent capacity factor, and the waste fuel consumption ranges between 434 tons/day (103,000 tons/year) for tire-derived fuel and 1177 tons/day (279,000 tons/year) for wood waste, based on 20 percent heat input from the waste fuel. Sewage sludge is cofired at 2 percent of total heat input, and the annual sludge consumption is 1335 tons/day (317,000 tons/year). The dedicated waste-to-energy facilities are sized to consume the same annual quantities of waste fuels, while operating at 80 to 85 percent capacity factor. As shown in Table II, net generating capacities range between 22 and 27 MW, and waste fuel consumption varies between 332 tons/day (103,000 tons/year) for whole tires and 900 tons/day (279,000 tons/year) for waste wood. The MSW-fired mass burn plant consumes 927 tons/day (288,000 tons/year) of unprocessed MSW.

Total Capital Requirement. Total capital requirement includes direct and indirect field erection costs, as well as the costs of engineering and home office services, project and process contingencies, escalation, interest during construction, preproduction and startup, inventory, and land. Table II and Figure 5 illustrate that total capital requirements for waste fuel energy conversion are significantly lower for the waste-fuel/coal cofired plants ($900 to $22,000/ton/day) than for the dedicated waste-to-energy plants ($54,000 to $258,000/ton/day).

Breakeven Fuel Price. The breakeven fuel price for the waste fuel-cofired plants is the price that results in no change in the cost of power generation relative to the unconverted coal-fired plant. The breakeven fuel price is $0.74/ton for tire-derived fuel cofiring, and is negative for the other fuels (-1.37 to -4.44 $/ton), i.e., the utility is paid to take the fuel. This occurs, because the incremental capital and O&M charges exceed the coal savings for RDF and wood cofiring. Sewage sludge cofiring actually increases coal consumption due to its large negative impact on boiler efficiency, and the fuel credit therefore decreases with increasing coal price.

Breakeven Tipping Fee. The breakeven tipping fee represents the charge for waste disposal required to balance total processing costs and total revenues derived from tipping fees and sale of recovered energy and byproducts. As shown in Table II and Figure 6, at $1.60/MBtu coal purchase price, the breakeven tipping fee for the waste fuel-cofired plants ranges between $11.40 and $41.65/ton waste. For the dedicated plants, at $0.05/kWh electricity sale price, the breakeven tipping fee ranges between $8.80 and $115.00/ton waste. Because these breakeven tipping fee estimates are quite sensitive to the assumed coal and electricity prices, the tipping fee ranges were also estimated for coal prices between $0.50 and 2.50/MBtu and for electricity prices between $0.03 and $0.10/kWh (Table II). Even for the wide ranges of coal and electricity prices, the cofired plants offer generally lower breakeven tipping fees than the dedicated plants. Note that the breakeven tipping fee range in Table II is reversed for sewage sludge cofiring ($6 to 2/ton) due to the inverse relationship between the fuel credit and coal price described in the previous paragraph.

Discussion

Although the estimated breakeven tipping fees are lowest for waste fuel cofiring in utility boilers, there are several institutional and economic factors that create barriers to implementing such projects. Regulated utilities typically pass on fuel savings to the rate payer as part of the rate making process and thus do not share directly in the economic benefits of waste fuel cofiring. Thus there is little incentive for a utility to participate other than to provide a service to the community and reduce landfill requirements. Economic dispatch of the utility system may also limit the hours when the power plant is available to consume waste to the point that the plant must operate at a higher rate and incur an economic dispatch penalty in order to consume all of the waste. Other important factors include the uncertainties created by potential environmental emissions from the waste fuels, and the separate and sometimes conflicting interests of the utilities, waste haulers, and municipalities. Clearly, a mechanism needs to be developed to share the financial and other risks as well as the economic benefits of waste fuel cofiring among all participants.

Conclusions

* Significant and growing quantities of alternate fuels are available for partial replacement of coal in steam and power generation.

* Cofiring waste fuels with coal in retrofitted coal-fired power plants and other industrial boilers offers the potential for higher energy recovery efficiency and lower breakeven waste tipping fees than dedicated waste-to-energy plants.

* Institutional constraints may limited cofiring of waste fuels in the future, unless the fuel supplier and user develop a mechanism to share the profits as well as the risks.

Literature Cited

1. Hughes, Evan E., "An Overview of EPRI Research on Waste-to-Energy," *Proceedings: 1989 Conference on Municipal Solid Waste as Utility Fuel*, EPRI GS-6994, February **1991**, pp.1-9 to 1-34.
2. *Alternative Fuel Firing in an Atmospheric Fluidized-Bed Combustion Boiler*, EPRI CS-4023, June **1985**.
3. Murphy, Michael L., "Fluidized Bed Combustion of Rubber Tire Chips: Demonstration of the Technical and Environmental Feasibility", presented at the IGT/CBETS Energy from Biomass and Wastes XI Conference, Lake Buena Vista, Florida, March **1987**.
4. Howe, W. C. and C. R. McGowin, "Fluidized Bed Combustion of Alternate Fuels: Pilot and Commercial Plant Experience," *Proceedings: 1991 International Conference on Fluidized Bed Combustion*, ASME, Montreal, Canada, April **1991**, pp. 935-946.
5. *Waste-to-Energy Screening Guide*, draft final report, EPRI Project 2190-5, January **1991**.

6. Makansi, Jason, "Ebara internally circulating fluidized bed (ICFB) technology", *Power*, January **1990**, pp 75-76.

7. McGowin, C. R., "Guidelines for Cofiring Refuse-Derived Fuel in Electric Utility Boilers", *Proceedings: 1989 Conference on Municipal Solid Waste as a Utility Fuel*, EPRI GS-6994, February **1991**, p. 2-47 to 2-66.

8. Zylkowski, Jerome R. and Rudy J. Schmidt, "Waste Fuel Firing in Atmospheric Fluidized Bed Retrofit Boilers," *Proceedings: 1988 Seminar on Fluidized Bed Technology for Utility Applications*, EPRI GS-6118, February **1989**, p. 2-47 to 2-61.

9. Coleville, Erik E. and Patrick D. McCarty, "Repowering of the Tacoma Steam Plant No. 2 with Fluidized Bed Combustors Fired on RDF, Wood, and Coal," presented at *Power-Gen '88 Conference*, Orlando, Florida, December **1988**.

10. Follett, R. E. and M. J. Fritsch, "Two Years of RDF Firing in Converted Stoker Boilers," *Proceedings: 1989 Conference on Municipal Solid Waste as a Utility Fuel*, EPRI GS-6994, February **1991**, pp. 3-25 to 3-44.

11. *Proceedings: 1991 Conference on Waste Tires as a Utility Fuel*, EPRI GS-7538, September **1991**.

12. "Hurricane Hugo Wood Chips Making Electricity," *The Logger and Lumberman*, May **1990**.

13. McGowin,C. R., "Alternate Fuel Cofiring in Utility Boilers", *Conference Proceedings: Waste Tires as a Utility Fuel*, EPRI GS-7538, September **1991**.

14. *Guidelines for Cofiring Refuse-Derived Fuel in Electric Utility Boilers, Vol. 2: Engineering Evaluation Guidelines*, EPRI CS-5754, June **1988**.

15. McGowin, C. R. and E. E. Hughes, "Coal and Waste Fuel Cofiring in Industrial and Utility Applications", *Proceedings: Eighth Annual International Pittsburgh Coal Conference*, Pittsburgh, PA, October 14-18, **1991**, pp. 853-858, p. 93.

16. Hollenbacher, Ralph, "Biomass Combustion Technologies in the United States", presented at the USDOE/EPRI/NREL Biomass Combustion Conference, Reno, Nevada, January **1992**.

17. "Wheelabrator Environmental Systems, Inc. Refuse-to-Energy System", Wheelabrator Environmental Systems, Hampton, NH, January, **1992**.

RECEIVED August 14, 1992

MUNICIPAL SOLID WASTE AND BIOMASS

Chapter 3

Recovery of Ethanol
from Municipal Solid Waste

M. D. Ackerson, E. C. Clausen, and J. L. Gaddy

Department of Chemical Engineering, University of Arkansas,
Fayetteville, AR 72701

Methods for disposal of MSW that reduce the
potential for groundwater or air pollution will be
essential in the near future. Seventy percent of
MSW consists of paper, food waste, yard waste, wood
and textiles. These lignocellulosic components may
be hydrolyzed to sugars with mineral acids, and the
sugars may be subsequently fermented to ethanol or
other industrial chemicals. This chapter presents
data on the hydrolysis of the lignocellulosic
fraction of MSW with concentrated HCl and the
fermentation of the sugars to ethanol. Yields,
kinetics, and rates are presented and discussed.
Design and economic projections for a commercial
facility to produce 20 MM gallons of ethanol per
year are developed. Novel concepts to enhance the
economics are discussed.

The United States generates about 200 million tons of MSW
annually, or about 4 pounds per capita per day (1). The average
composition of MSW is given in Table I, and varies slightly with
the season (2). This material has traditionally been disposed of
in landfills. However, recent environmental concerns over ground
water pollution, leaching into waterways, and even air pollution,
as well as increasing costs, have resulted in this technology
becoming unacceptable in most areas. Few new landfills are being
approved, and the average remaining life of operating landfills is
only about five years.
  Alternatives to landfilling include incineration, composting,
anaerobic digestion, and recycling. Incineration can result in
energy recovery as steam. However, concerns over hazardous
components in exhaust gases and high capital and operating costs
detract from this alternative. Large areas required for
composting and the ultimate use or disposal of compost with high
metals content makes this technology uncertain. Very slow
reaction rates and large reactors for anaerobic digestion makes
this technology uneconomical at present.

0097–6156/93/0515–0028$06.00/0

Table I. Municipal Solid Waste Composition
(Weight Percent as Discarded)

Category	Summer	Fall	Winter	Spring	Average
Paper	31.0	38.9	42.2	36.5	37.4
Yard Waste	27.1	6.2	0.4	14.4	13.9
Glass	17.7	22.7	24.1	20.8	20.0
Metal	7.5	9.6	10.2	8.8	9.8
Wood	7.0	9.1	9.7	8.2	8.4
Textiles	2.6	3.4	3.6	3.1	3.1
Leather & Rubber	1.8	2.5	2.7	2.2	2.2
Plastics	1.1	1.4	1.5	1.2	1.2
Miscellaneous	3.1	4.0	4.2	3.7	3.4

Recycling of glass, metals, plastics, and paper reduces the
quantity of material to be landfilled by about 60 percent, as seen
from Table I. Most states have decided that recycling offers the
best solution to the environmental concerns associated with solid
waste disposal and many have implemented regulations for curbside
segregation of recyclable components. Markets for recycled
aluminum and steel are well established, however, markets for
recycled paper, glass, and plastics are not well developed. Low
prices (negative in some areas for paper) will impede the
application of recycling.

Alcohol Production. The United States currently imports about
half of its crude oil and must produce another 120 billion gallons
of liquid fuels annually to become energy self sufficient.
Ethanol can be produced from lignocellulosic matter, like paper,
by hydrolysis of the polysaccharides to sugars, which can be
fermented into ethanol. This technology would enable the use of
the entire carbohydrate fraction of MSW (paper, yard and food
waste, wood and textiles), which constitutes 75 percent of the
total, into a useful and valuable product. Ethanol can be blended
with gasoline and, currently, nearly one billion gallons of
ethanol, primarily made from corn, are used as a transportation
fuel in this country. The potential market (at 10 percent
alcohol) is 10 billion gallons per year. Blending of ethanol with
gasoline reduces emissions and increases the octane rating. Some
states, like California and Colorado where air quality has
degraded seriously in metropolitan areas, are mandating the use of
alcohol fuels.
    The purpose of this paper is to describe a process for
converting the lignocellulosic fraction of MSW into ethanol. The
residue is contacted with concentrated mineral acid at room
temperature to give theoretical yields of monomeric sugars, which
are readily fermented into ethanol. Procedures to give high sugar
concentrations are described. Data for fermentation in
immobilized cell columns in a few minutes are presented. The
economics of this process is then developed and key economic
parameters identified.

HYDROLYSIS/ETHANOL PRODUCTION

The hydrolysis of biomass to sugars and fermentation of glucose to
ethanol are technologies that have been commercial around the
world for many years. The U. S. produced up to 600 million
gallons of ethanol per year by fermentation during World War II.
Also, the Germans produced fuel ethanol from wood by hydrolysis
and fermentation during World War II. Today, Brazil produces most
of its liquid fuel from sugar cane.
     It has been known for nearly two centuries that cellulose
could be converted into sugars by the action of mineral acids (3).
The process became commercial early in this century with dilute
acid plants built in Georgetown, South Carolina, and Fullerton,
Louisiana to produce 2-3 million gallons of ethanol per year from
wood (4). These plants operated through the end of World War I
when wood sugars could not complete with cheap by-product molasses
from cane.
     During World War II, the Germans developed a percolation
process and built 20 plants for the production of fuel alcohol
from wood (5). Similar plants were also built in Switzerland,
Sweden, China, Russia, and Korea. In attempts to produce ethanol
for butadiene rubber production, the United States built a 4
million gallon per year wood hydrolysis to ethanol facility in
Springfield, Oregon in 1944. The Germans also operated plants
based upon the Bergius concentrated hydrochloric acid technology
at Mannheim and Regensburg during World War II (6). Concentrated
sulfuric acid plants were also operated in Italy (Giordani Leone)
and Japan (Hokkaido) (7). Most of these facilities were closed
after World War II with the development of processes to produce
ethanol from petroleum. However, about 40 percolation plants are
still operated in Russia today.

Hydrolysis Technology. Biomass materials are comprised of three
major components: hemicellulose, cellulose, and lignin. The
composition of various biomass materials is shown in Table II. As
noted, most of these materials contain 50-70 percent carbohydrate
(hemicellulose and cellulose). These polysaccharides can be
hydrolyzed to monomeric sugars, which can be converted by
microorganisms into fuels or chemicals. The lignin cannot be
hydrolyzed, but has a high heating value and can be used as a
source of fuel. From Table II, most of the MSW biomass is
cellulose.

Table II. The Composition of Selected Biomass Materials

Material	Percent Dry Weight of Material		
	Hemicellulose	Cellulose	Lignin
Tanbark Oak	19.6	44.8	24.8
Corn Stover	28.1	36.5	10.4
Red Clover Hay	20.6	36.7	15.1
Bagasse	20.4	41.3	14.9
Oat Hulls	20.5	33.7	13.5
Newspaper	16.0	61.0	21.0
Processed MSW	25.0	47.0	12.0

The carbohydrate hydrolysis can be carried out by contact with cellulase or xylanase enzymes, or by treatment with mineral acids. Enzymatic hydrolysis has the advantage of operating at mild conditions and producing a high-quality sugar product. However, the enzymatic reactions are quite slow (30 hour retention time), and the biomass must be pretreated with caustic or acid to improve the yields and kinetics. The expense of pretreatment and enzyme production, and the large reactors required make this an uneconomical alternative.

Acid hydrolysis is a much more rapid reaction and various combinations of temperature and acid concentration may be used. Two methods of acid hydrolysis have been studied and developed: a high temperature, dilute acid process (8,9) and a low temperature, concentrated acid process (10,11). For example, complete conversion of the hemicellulose and cellulose in corn stover into monomeric sugars and sugar degradation products requires mineral acid concentrations of 2N at temperatures of 100-200°C (12). However, acid concentrations of 10-14N yield complete conversions at room temperature (30°C).

At high temperatures, xylose degrades to furfural and glucose degrades to 5-hydroxymethyl furfural (HMF), which are both toxic to microorganisms. Yields from dilute acid processes are typically only 50-60 percent of theoretical because of sugar losses by degradation and reverse polymerization at high temperatures. Also, equipment corrosion at high temperatures is a serious problem. Work in our laboratories has focused attention on concentrated acid processes which produce theoretical yields at low temperatures. However, since high acid concentrations are used, acid recovery is an economic necessity (10).

Studies in our laboratories have resulted in both single step and two-step hydrolysis processes, using concentrated mineral acids, which result in nearly 100 percent yields of sugars from hemicellulose and cellulose. The reactions are conducted at room temperature, without significant degradation or reverse polymerization (11-13). An acid recovery process has been developed and tested, yielding an energy efficient method of separating sugar and acid (14). The resulting sugar solution has been successfully fermented to ethanol and other chemicals without pretreatment (15).

Process Description. Figure 1 shows the proposed process for the acid hydrolysis of MSW consisting simply of a mixed reactor where acid and MSW are contacted at a constant temperature. The unconverted solids (lignin and ash) are separated by filtration, washed, and used as fuel. Acid and sugars are separated and the acid returned to the reactor.

If desirable to separate the sugars, the hemicellulose, which degrades at milder conditions, may be first hydrolyzed to produce a mixture of five and six carbon sugars. The solids from this first stage reactor are separated and contacted with acid in a second step to hydrolyze the cellulose. Only six carbon sugars are obtained from cellulose in this second stage. This two step hydrolysis gives two streams; a xylose rich prehydrolyzate and a glucose rich hydrolyzate; and may be used where sugar separation

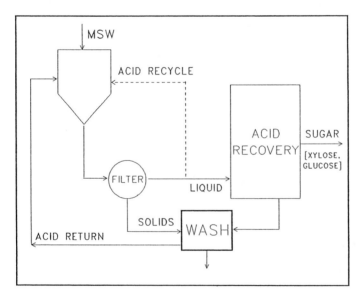

Figure 1. Schematic of Acid Hydrolysis. (Reproduced with permission from Energy from Biomass and Waste XV, Institute of Gas Technology, 1991. Copyright 1991 Michael D. Ackerson.)

is desirable. In the usual case, as with MSW, the simpler single step process will be preferred.

Hydrolysis Conditions. The two major factors which control the hydrolysis reactions are temperature and acid concentration. Studies in our laboratories have been made to define the appropriate conditions to maximize reaction rates and yields. Sugar degradation is promoted more at high temperature than at high acid concentration. Also, fast rates of hydrolysis are achieved at acid concentrations exceeding 12N. Therefore, the best conditions are a high acid concentration (80 percent H_2SO_4 or 41 percent HCl) and a mild temperature (~40°C).

The sugar concentrations and yields from a typical hydrolysis of MSW from our laboratories are given in Table III (11). The prehydrolysis step yields 8 percent of the initial MSW as xylose. The combined yield of glucose is 60 percent. These yields represent nearly complete conversion of hemicellulose and cellulose to sugars. However, very dilute (3-7 percent) sugar solutions result from these reactions.

Feedstock Preparation. In order to speed up the hydrolysis reactions, the size of the biomass particles must be reduced to increase the accessibility to the polymeric structure. A high solids concentration is desirable since this concentration controls the sugar concentration and the size of the hydrolysis and fermentation equipment. The size of the particles also affects the fluidity of the solids/acid slurry. It is desirable to maintain fluidity of the slurry to promote mass transfer and to facilitate pumping and mixing. Therefore, the particle size is an important variable in the biomass conversion process.

Table III. MSW Acid Hydrolyzates

	Concentration g/L	Yield g/100g
Prehydrolyzate		
Xylose	9.5	8.0
Glucose	18.5	16.0
Hydrolyzate		
Xylose	0.0	0.0
Glucose	67.8	44.0
Combined		
Xylose		8.0
Glucose		60.0

Table IV gives the maximum solids concentration to maintain fluidity of the slurry, as a function of particle size. A maximum concentration of about 10 percent is possible with particle sizes less than 40 mesh. Grinding to 20 mesh gives a particle size distribution in which 90 percent of the material is less than 40 mesh. Therefore, grinding biomass to pass 20 mesh gives the appropriate size and produces the maximum possible slurry concentration. Also, grinding to smaller sizes does not improve the reaction rate.

Table IV. Maximum Solids Concentration for Fluid Slurry

Mesh Range (Sieve Nos.)	Solids Concentration wt. %
0 - 1	4.6
12 - 20	4.6
20 - 30	8.4
30 - 40	8.2
40 - 45	10.2
45 - 70	10.4
70 - 100	10.6
100+	10.8

Sugar Decomposition. The fermentability of the sugars is dependent upon the sugar decomposition that occurs during hydrolysis. Xylose decomposes to furfural and hexoses decompose to HMF, which are both toxic to yeast. Tolerance can often be developed, and toxicity is difficult to define. However, the toxic limit of furfural on alcohol yeast is reported to be 0.03 to 0.046 percent (16). HMF is reported to inhibit yeast growth at 0.5 percent, and alcohol production is inhibited at 0.2 percent (17).

The rate of decomposition of xylose to furfural and hexoses to HMF have been studied at varying sugar concentrations. Using the method of initial rates, these reactions were found to be first-order. The ratio of rate constants for decomposition to formation are given in Table V. These ratios are small, and subsequent calculations and experiments show that the rate of HMF appearance is insignificant. However, the rate of furfural appearance could reach toxic limits, especially if acid recycle is utilized.

Table V. Ratio of First Order Rate Constants for Sugar Decomposition to Formation Under Prehydrolysis Conditions

Sugar	Acid Concentration	Rate of Formation/Decomposition
Glucose	2N	0.0053
	3N	0.0090
	4N	0.0074
Xylose	2N	0.0257
	3N	0.0402
	4N	0.0374

Hydrolyzate Fermentation/Ethanol Production

Glucose may be fermented to ethanol efficiently by the yeast *Saccharomyces cerevisiae,* or the bacterium

Zymomonas mobilis (18). Batch fermentation experiments were
carried out to compare the production rates of ethanol from
hydrolyzates and synthetic glucose. *Saccharomyces cerevisiae*
(ATCC 24860) was used in the study. As shown in Figure 2,
identical results were found when fermenting synthetic glucose and
hydrolyzate. Ethanol yields were also nearly identical. As noted
in Table VI, the fermentation proceeded well in the presence of a
small amount (0.25 percent) yeast extract, which can be obtained
by recycle. Almost total conversion of sugars is obtained in only
16 hours. The concentrations of furfural and HMF in the
hydrolyzates were found to be negligible. These low levels of
byproducts are believed to be the major reason for this highly
efficient fermentation.

Table VI. Hydrolyzate Fermentation to Ethanol
Percent Sugar Utilization

	Hydrolyzate			
	With Vitamins and With Yeast Extract			
Fermantation Time (hrs)	Amino Acids	NH3PO3)	Amino Acids and NH3(PO3)	Yeast Extract
16	15.9	21.9	27.3	97.5
23	19.3	24.9	35.8	97.5

Xylose fermentation is much more difficult, and the xylose
might be used as a source of energy for generating steam and
power. However, future possibilities for xylose fermentation will
improve the economics. Recent work with *Pachysolen tannophilus*
shows promise for xylose conversion to ethanol (19) but, at
present, this technology is not fully developed. Ethanol may also
be produced by converting xylose to xylulose, followed by
fermentation with yeast (20).

Continuous Fermentation. The standard technology for fermenting
sugars to ethanol is in batch vessels. Batch fermentation is used
so that contamination and mutation can be controlled.
Sterilization between batches and the use of a fresh inoculum
insure efficient fermentation. However, most batch alcohol
fermentations are designed for thirty hour (or more) reaction
time, which results in very large and expensive reactors.
The reactor size can be reduced substantially by using
continuous flow fermenters. When fermenting acid hydrolyzates,
the problems with maintaining sterile conditions are substantially
reduced, since the substrate has been sterilized by contact with
the acid. Therefore, the use of continuous fermentation is a
natural application for producing alcohol from MSW hydrolyzates.
A number of continuous fermentation schemes have been studied,
including the CSTR (21), cell recycle reactor (22), flash
fermentation (23) and immobilized cell reactors (24,25).
Immobilized cell reactors (ICR) show potential in substantially

decreasing reactor size and decreasing substrate and product
inhibition (25-28). Reaction rates for ethanol production in an
immobilized cell reactor are as high as 10 times the values
obtained in a stirred tank reactor (24). A wide variety of
immobilization techniques have been employed, including cross-
linking, entrapment, and covalent bonding (25).

Data are given in Figure 3 for laboratory columns with
immobilized *S. cerevisiae*. The glucose profiles are given for
initial sugar concentrations from 50-200 g/L. As noted, 90
percent conversion is achieved in one hour or less.
Productivities to achieve 99 percent conversion were about 40 g/L-
hr, or about an order of magnitude greater than the CSTR and 60
times more than the batch reactor. Furthermore, alcohol
inhibition and toxicity to either inhibitors is reduced in the
ICR. The volume of the ICR for MSW hydrolyzate fermentation is
about 5 percent that of the batch fermenter and substantial
capital savings result.

Increasing the Sugar Concentration

Perhaps the single most important factor in the economics of this
process is the sugar concentration that results from acid
hydrolysis. Dilute concentrations increase both the equipment
size and the energy required for purification. Methods to
increase the sugar and ethanol concentrations have been developed.

Solids Concentration. The ultimate sugar and alcohol
concentrations are direct functions of the initial solids
concentration in the hydrolysis. Since fluidity in a stirred
reactor is a requirement, a 10 percent mixture has been considered
maximum. Therefore, the resultant sugar concentrations have been
only 2-7 percent and alcohol concentrations only half as much.

If the limiting factor is considered to be fluidity in the
reactor, instead of the feed mixture, the feed concentration could
be increased by roughly the reciprocal of one minus the solids
conversion in the reactor. Of course, solids and liquid would
have to be fed separately, which could also save equipment cost.
For biomass, containing 75 percent carbohydrate, the reactor size
could be reduced by 75 percent. Attendant reductions would also
result in the filtration and washing units.

Equally important are the resultant increases in sugar
concentrations. The glucose concentration would be quadrupled to
about 280 g/L (28 percent). Energy and equipment costs in the
fermentation area would be reduced proportionately. This simple
alteration in the process has a profound impact on the economics.
It is estimated that the capital cost reduced by 40 percent in the
hydrolysis and acid recovery sections and 60 percent in the
fermentation and utilities areas. Furthermore, the energy
requirements for distillation are reduced by 60 percent.

Acid Recycle. Another method to increase the sugar concentration
is to recycle a portion of the filtrate (acid and sugar solution)
in the hydrolysis step. The acid would catalyze further
polysaccharide hydrolysis to increase the sugar concentration. Of
course, recycle of the sugars would also increase the possible
degradation to furfural and HMF.

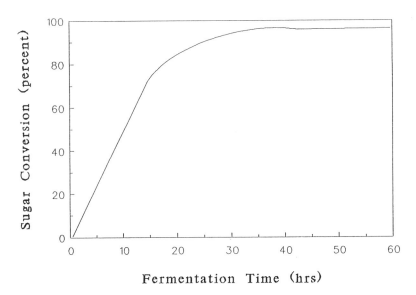

Figure 2. Fermentation of Hydrolyzate and Synthetic Glucose. (Reproduced with permission from Energy from Biomass and Waste XV, Institute of Gas Technology, 1991. Copyright 1991 Michael D. Ackerson.)

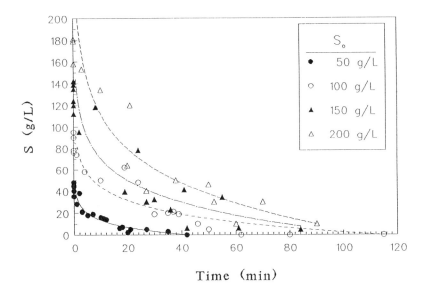

Figure 3. Glucose Profile in the ICR. (Reproduced with permission from Energy from Biomass and Waste XV, Institute of Gas Technology, 1991. Copyright 1991 Michael D. Ackerson.)

Experiments have been conducted to determine the enhancement possible with acid recycle. Various amounts of the acid and sugar solution from the filtration were recycled to determine the resulting sugar and by-product concentrations. Acid and solids concentrations and temperatures were kept constant. These experiments have shown that the sugar concentrations can be increased sixfold at total recycle. It should be noted that not all the filtrate can be recycled, since a portion adheres to the solids in filtration. In order to minimize sugar decomposition, a recycle fraction of 50 percent has been used, which results in doubling the sugar composition, without significant furfural or HMF levels.

The effect of acid recycle on the economics is significant. A recycle rate of 50 percent, coupled with high solids concentrations, would result in a xylose concentration of 15 percent and a glucose concentration of over 50 percent could be achieved. Practically, sugar concentrations should not exceed 25 percent, so a smaller recycle fraction is required. It should be noted that these concentrations have been achieved in the laboratory, while maintaining furfural and HMF less than 0.05 percent. These high concentrations reduce the equipment size in the acid recovery section by 50 percent and in the fermentation section by another 60 percent. Energy consumption is also reduced another 60 percent.

Acid Recovery. Acid recovery is essential when using concentrated acid hydrolysis. Processes for recovery of both hydrochloric and sulfuric acids have been developed. A number of possible recovery schemes were examined, including electrodialysis, distillation, etc.

The recovery technology that has been selected is based upon solvent extraction. Solvents have been identified that extract HCl and H_2SO_4 from the aqueous sugar solutions. Near complete acid recovery is possible and solvent losses are minimized. For HCl, the acid is separated from the solvent by distillation, and the solvent recycled. A hexane wash of the sugar solution is used to recover trace quantities of solvent, and hexane is separated by distillation for recycle.

Some solvent is lost in the process; however, the losses are quite small and solvent replacement costs are only $0.02 per gallon of alcohol. Acid losses are minimized and acid costs are $0.025 per gallon of alcohol. The total heat requirement for solvent and acid recovery is low and amounts to less than $0.05 per gallon of alcohol. As shown later, the energy cost may be recovered from the lignin and xylose streams.

ECONOMIC PROJECTIONS

To illustrate the economics of this process, a design has been performed for a facility to convert MSW into 20 million gallons per year of ethanol, utilizing the acid hydrolysis procedures previously described. The capital and operating costs are summarized in Table VII.

MSW would be collected and delivered to the plant site as

needed. Feedstock preparation consists of plastic, metal and glass removal, shredding, grinding and conveying to the reactors. The cost of the removal of glass and metals is not included in the feed processing cost, as reports indicate that resale of these materials will offset the capital and operating costs of separation. The hydrolysis section, as shown in Figure 1, consists of continuous reactors. Acid resistant materials of construction are necessary for this equipment. Ethanol fermentation in the ICR and typical distillation units are included. The total capital cost for this plant is $35 million, including all utilities, storage and offsite facilities.

The annual operating costs are also shown in Table VII. These costs are also given on the basis of unit production of alcohol. As mentioned previously, no cost is included for MSW. A lignin boiler is used to reduce the energy requirements, and energy costs are only $0.08 per gallon. Fixed charges are computed as a percentage of the capital investment and total $5.6 million per year. The present ethanol price of $1.50 per gallon will generate revenues of $30 million and yield a pre-tax profit of $18.5 per year ($.93/gal) or 53 percent per year.

It should be noted that this process does not include utilization of the pentose stream. Acid recovery is included, but fermentation of the xylose is not provided. Xylose could be fermented to alcohol, acids or other valuable chemicals, which would improve the economics. However, since this technology is not perfected, such products have not been included.

Table VII.   Economics of 20 Million Gallon Per Year
Ethanol Facility

A. Capital Cost

	Million $
Feedstock Preparation	3.0
Hydrolysis	5.0
Acid Recovery	8.5
Fermentation & Purification	8.0
Utilities/Offsites	6.5
Engineering	4.0
	35.0

B. Operating Cost

	Million $/yr	$/gal
MSW	-	-
Utilities	1.5	0.08
Chemicals	1.9	0.09
Labor	2.5	0.13
Fixed Charges		
Maintenance (4%)	1.4	0.07
Depreciation (10%)	3.5	0.18
Taxes & Insurance (2%)	0.7	0.02
Pre-tax Profit (53%)	18.5	0.93
	$30.0	$1.50/gal

Sensitivity analyses show that the economics are particularly sensitive to capital cost and revenue. A 20 percent reduction in capital cost raises the pre-tax return to 70 percent. Similarly, a 20 percent increase in the ethanol price increases the return to 70 percent. A tipping fee of $20 per ton of MSW would increase the return to 65 percent. Increasing the plant size would have a similar positive affect on the economics.

CONCLUSIONS

Concentrated acid hydrolysis of residues, such as MSW, requires mild temperatures and results in near theoretical yields. The resulting hydrolyzates, containing primarily xylose and glucose, can be fermented to ethanol or other chemicals. The acid can be recovered for reuse by solvent extraction. The sugar concentration can be increased by using high solids reactions with acid recycle. Continuous fermentation of the hydrolyzates can be achieved in an hour or less in an immobilized cell column. The capital cost for a process to produce 20 million gallons of ethanol is estimated to be $35 million. The pre-tax profit from this facility is sufficient to encourage commercialization.

LITERATURE CITED

1. Banerjee, N. and Vishwanathan, L., *Proc. Annual Sugar Conv. Tech. Conf.*, CA 62: 1228f (1972).
2. Banerjee, N. and Vishwanathan, L., *Proc. Annual Sugar Conv. Tech. Conf.*, CA 82: 122177W (1974).
3. Braconnet, H., *Ann. Chem. Phys.*, 12-2, p. 172 (1819).
4. Clausen, E. C. and Gaddy, J. L., "Production of Ethanol from Biomass," *Biochem. Engr.* (1982).
5. Clausen, E. C. and Gaddy, J. L., "Economic Analysis of Bioprocess to Produce Ethanol from Corn Stover," *Biotech. and Bioengr.*, 13 (1983).
6. Clausen, E. C. and Gaddy, J. L., "The Production of Fuels and Chemicals from MSW," *Biotechnology Applied to Environmental Problems*, CRC Press, (1985).
7. Clausen, E. C. and Gaddy, J. L., "Acid Hydrolysis/ Recovery," SERI Review Meeting, Golden, CO (October 1987).
8. Cysewski, G. R. and Wilke, C. R., *Biotech. Bioeng.*, 20, 1421 (1978).
9. Cysewski, G. R. and Wilke, C. R., Lawrence Berkeley Laboratory Report 4480, (March 1976) *Biotech. Bioeng.*, 19, 1125 (1977).
10. Elias, S., *Food Engineering*, p 61 (October 1979).
11. Faith, W. L., "Development of the Scholler Process in the United States," *Ind. and Engr. Chem.*, 37.1, p. 9 (1945).
12. Gainer, J. L. *et al.*, "Properties of Adsorbed and Covalently Bonded Microbes," presented at AIChE National Meeting, New Orleans (1981).
13. Goldstein, I. S., "Organic Chemicals from Biomass," CRC Press, Boca Raton, FL, p. 102 (1980).
14. Gong, C. S., McCracken, L. D., and Tsao, G. T., *Biotech Letters*, 3, 245-250 (1981).

15. Grethlein, H. E. and Converse, O. C., "Continuous Acid Hydrolysis for glucose and Xylose Fermentation," presented at International Symposium on Ethanol from Biomass, Canertech Ltd, Winnipeg (1982).
16. Hall, J. A., Saeman, J. F., and Harris, J. F., "Wood Saccharification," *Unasylva*, 10.1, p. 7 (1956).
17. Linko, P., "Immobilized Microbial Cells for Ethanol and Other Applications," presented at AIChE National Meeting, New Orleans (1981).
18. Ng, A. S., Wong, M. K. Stenstrom, Larson, L., and Mah, R. A., "Bioconversion of Classified Municipal Solid Wastes: State-of-the-Art Review and Recent Advances," *Fuel Gas Developments*, CRC Press, (1983).
19. Prieto, S., Clausen, E. C. and Gaddy, J. L., "The Kinetics of Single-Stage Concentrated Sulfuric Acid Hydrolysis," *Proceedings Energy from Biomass and Wastes XII*, (1988a).
20. Prieto, S., Clausen, E. C., and Gaddy, J. L., "Ethanol from Biomass by Concentrated ACid Hydrolysis and Fermentation," *Proceedings Energy from Biomass and Wastes XII*, (1988b).
21. Rowe, G. and Magaritis, A, "Continuous Ethanol Production in a Fixed-Bed Reactor Using Immobilized Cell of *Zymomonas mobilis*," presented at AIChE National Meeting, New Orleans, (1981).
22. Rugg, B., "Optimization of the NYU Continuous Cellulose Hydrolysis Process," SERI Report No. TR-1-9386-1, (1982).
23. Schneider, H., Yang, P. Y., Chan, Y. K., and R. Maleszka, *Biotech. Letters*, 3, 89-92 (1981).
24. Sherrod, E. C. and Kressman, F. W., "Sugars from Wood: Review of Processes in the United States Prior to World War II," *Ind. and Eng. Chem.*, 37.1, p 4 (1945).
25. Sitton, O. C. and Gaddy, J. L., *Biotech. Bioeng.*, 22, 1735, (1980).
26. Vega, J. L., Clausen, E. C., and Gaddy, J. L., "Biofilm Reactors for Ethanol Production," *Enzyme and Microbial Tech.*, 10(3), 389 (1988).
27. Waldron, R. D., Vega, J. L., Clausen, E. C., and Gaddy, J. L., "Ethanol Production Using *Zymomonas mobilis* is Cross-Linked Immobilized Cell Reactors," *Appl. Biochem. and Biotech.*, 18, 363 (1988).
28. _____ U. S. E. P. A., "Technical Environmental and Economic Evaluation of Wet Processing for Recovery and Disposal of Municipal Solid Waste," SW-109C (1981).

RECEIVED October 19, 1992

Chapter 4

Converting Waste to Ethanol and Electricity via Dilute Sulfuric Acid Hydrolysis

A Review

L. G. Softley, J. D. Broder, R. C. Strickland, M. J. Beck, and J. W. Barrier

Biotechnical Research Department, Tennessee Valley Authority,
Muscle Shoals, AL 35660

In April 1990, TVA began a project to evaluate the processing of municipal solid waste (MSW) to recyclables, ethanol, and electricity. The project includes evaluation of front-end classification processes for recovery of recyclables, hydrolysis and fermentation of the cellulosic fraction of MSW to ethanol, and combustion of the hydrolysis residue for steam and electricity production. Laboratory hydrolysis and fermentation tests have resulted in yields of up to 36 gallons of ethanol per dry ton of the cellulosic fraction of MSW. Yields of up to 29 gallons per dry ton have been attained in pilot plant runs. Analyses of process effluents have been performed to evaluate the environmental acceptability of the overall process. A preliminary economic evaluation has been conducted based on these tests.

Each year about 240 million tons of municipal solid waste (MSW) are generated in the United States. Of this amount, about 160 million tons are generated by consumers in the form of residential and commercial waste. This figure is expected to reach 193 million tons by the year 2000 (1). While MSW generation is increasing, the number of landfills is dropping significantly. According to the Environmental Protection Agency (EPA), more than half of our existing landfills will reach their capacity within the next eight years. Many landfills will be forced to close because of tougher environmental regulations and because of potential ground-water contamination from leaking and leaching.

As a result of these problems, several disposal alternatives have been developed to reduce the amount of MSW requiring landfilling. These alternatives to landfilling include incineration/mass burn (burning the entire waste stream) and refuse-derived fuel (RDF) combustion (burning the waste after recyclables have been removed). About 13% of the United States' waste is treated in this manner (2). Both of these alternatives produce steam and electricity, but are expensive and environmentally questionable. In response to the nation's waste problem, the

0097–6156/93/0515–0042$06.00/0

Tennessee Valley Authority (TVA) is developing technology to produce recyclables, electricity, ethanol, and other chemicals from MSW.

The National Fertilizer and Environmental Research Center (NFERC) at TVA has conducted laboratory hydrolysis and fermentation research to convert cellulosic biomass materials to ethanol and other chemicals since 1980. In 1985, a 2-ton-per-day pilot plant was built to convert hardwoods to ethanol and other chemicals. Previous TVA research was designed to maximize sugar yield from both hemicellulose and cellulose in a two-stage process. Because of the costs associated with hemicellulosic sugar conversion and the small quantity of pentose sugars associated with MSW cellulosics, conversion by a one-step hydrolysis of cellulose is being evaluated for producing hexose sugars. In contrast to previous work on hardwood and agricultural residue conversion, there are important environmental questions which must be answered before an economical large-scale wastepaper or MSW-cellulosics conversion facility is possible. Recently, TVA conducted preliminary laboratory and pilot-plant tests with MSW and waste paper to evaluate their potential and also to address environmental concerns related to these feedstocks. Based on the results of these tests, a preliminary technical and economic evaluation was performed.

In April 1990, TVA began research to evaluate conversion of MSW-cellulosics to ethanol and electricity. This paper presents a process description, laboratory and pilot plant results, and safety and environmental concerns. TVA's integrated processing system produces recyclables, ethanol, chemicals, and electric power from MSW. This system is divided into three areas: MSW classification, RDF processing, and energy production. A schematic of the system is shown in Figure 1. The first step involves classification of MSW to remove the recyclables (glass, plastics, aluminum, steel, and other metals, etc.). Cardboard, newsprint, and paper products are left in the waste stream (RDF) as potential feedstocks for chemical production.

The second step involves RDF processing. Dilute sulfuric acid hydrolysis is used to convert the RDF to fermentable sugars for ethanol production. A flow diagram of the hydrolysis and fermentation process is shown in Figure 2. RDF is fed to the hydrolysis reactor where dilute sulfuric acid (2-3%) is added. The mixture is then steam heated to 160-190° C with retention times ranging from 5 to 30 minutes. Under these conditions, the cellulose and hemicellulose are converted to sugars (glucose and xylose/mannose respectively). During hydrolysis, small quantities of acetic acid and sugar degradation products such as furfural and hydroxymethyl furfural are also formed. The mixture is dewatered, and the remaining solids are used for boiler fuel. The sulfuric acid in the solution is neutralized with lime, and the resulting mixture is filtered to remove the gypsum. The sugar solution is fermented to ethanol. Beer from fermentation is distilled, and the ethanol dehydrated for use as a motor fuel additive(*3*).

The third step in the process is energy production. This involves combustion of the solid residue remaining after hydrolysis. About 40% of the solids that enter the hydrolysis process are removed as residue. This solid residue contains lignin (about 15-25% of the original feedstock), unreacted cellulose, and ash, and has an energy content of about 8,500 Btu/pound. These solids are fed to a boiler where

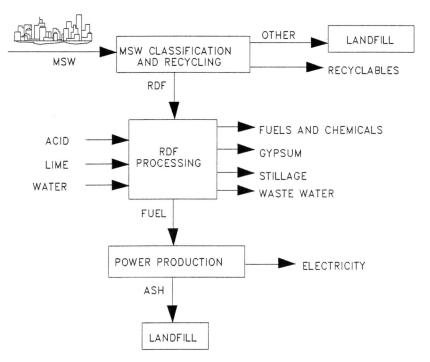

Figure 1.   Integrated MSW Processing System.

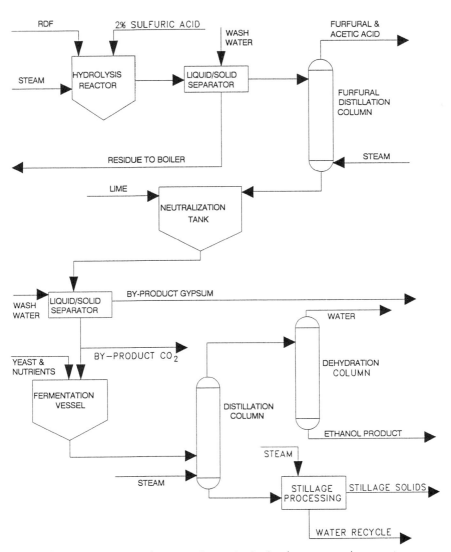

Figure 2.   Flow diagram of RDF hydrolysis processing system.

process steam and electricity are produced. More electricity is produced than is needed in the process, and it can be sold as a by-product.

Laboratory And Pilot Plant Results

A series of dilute sulfuric acid hydrolysis tests were conducted using shredded newsprint, cardboard pellets, newsprint pellets, and RDF pellets as feedstocks. The shredded newsprint was newsprint that had the glossy and highly-colored sections removed. The pelletized feedstocks were received from a commercial classification facility in Humboldt, Tennessee. The pellets were approximately 0.5 inches in diameter and of variable lengths of up to three inches.

Laboratory hydrolysis tests were conducted at 130-190° C in an acid-resistant, Monel-lined, 20-liter, rotating digester using either 1% (w/w), 2% (w/w), or 4% (w/w) sulfuric acid in a one-stage dilute sulfuric acid hydrolysis. In preliminary tests, samples of the hydrolysis liquid were collected at 5-to-10 minute intervals for 3 to 4 hours. Additional tests were performed where samples were collected at 1-minute intervals. Tests were performed to evaluate the effect of liquid-to-solid ratio (liquid-to-solid ratios adjusted for moisture content) on sugar production. It has been determined that a 5:1 liquid-to-solid ratio was adequate to maintain free liquid to saturate the feedstock. Several additional tests were performed where the residues were hydrolyzed a second time to evaluate increases in sugar yields from the cellulose remaining from the one-step hydrolysis. Total reducing sugars were measured by the dinitrosalicylic (DNS) method and selected samples were analyzed by HPLC. Residues were analyzed for moisture content, weight loss, Btu content, and compositional changes. Inorganic analyses were conducted by TVA's Chemical Environmental Analysis Section.

Laboratory Results. Previous work by Strickland et al., 1990, (4), Broder et al., 1991, (5), Broder et al., 1991, (6), and Barrier et al., 1991, (7) discussed results of TVA laboratory tests using waste cellulosics as feedstock. Table I summarizes yield information from MSW-cellulosics and newsprint hydrolysis in the laboratory. Table I also shows improvements in sugar yield where the residues were hydrolyzed a second time.

Laboratory hydrolysis tests have produced yields of up to 20 gallons per dry ton of RDF with a single-stage process and 27 gallons per ton with a two-stage process. Newsprint hydrolysis tests have produced yields of up to 24 gallons per dry ton with a single stage process and 44 gallons per dry ton with a two-stage process (5). Additional tests at higher temperatures are shown in Table II.

The tests in Table II were different from previous hydrolysis tests because power to the heating element in the digester was cut off at 160-165° C and the heat inside the digester continued to rise to 200-205° C (maximum temperature for previous tests has been 130-190° C). In previous tests, the objective had been to reach and maintain a specified temperature. The tests reported in Table II were conducted after reaching the maximum temperature for the digester as quickly as possible with hydrolyzate samples being collected along this temperature profile. Due to an increase in pressure (over 20 atmospheric barrs), all tests did not reach

Table I. Laboratory sugar yields from MSW-cellulosics and newsprint[a]

	Time (min)[b]	Temp (C)	Max Sugar Concentration (%)	Ethanol Yield (gal/t)[c]
ONE-STAGE HYDROLYSIS				
Humboldt-160	15	160	1.94	15.0
Humboldt-170	12	170	2.18	16.6
Humboldt-180	-42	178	2.25	17.4
Humboldt-190	-34	180	2.65	20.5
Newsprint-150	120	150	2.65	20.5
Newsprint-160	15	160	3.01	23.8
Newsprint-170	15	170	2.69	20.8
Newsprint-180	0	180	2.86	22.1
TWO-STAGE HYDROLYSIS				
Humboldt-one	15	160	1.94	15.0
Humboldt-two	25	170	2.32	12.0
TOTAL[d]				27.0
Newsprint-one	15	160	3.01	23.8
Newsprint-two	30	170	3.94	20.5
TOTAL				44.3

[a] 5:1 liquid-to-solid ratio and 2% sulfuric acid
[b] Positive numbers indicate hydrolysis time after reaching target temperature; negative numbers indicate time from start of heat-up. (35-45 minutes were required to reach the desired temperature after starting the tests.)
[c] Assumes 100% sugar recovery, 100% sugar conversion, theoretical ethanol yield, and 100% ethanol recovery.
[d] Yield from one ton of feedstock.

Table II. Results of high temperature hydrolysis tests in the rotating digester [a]

Feedstock[b]	Time[c] (min)	Temp (C)	Glucose (lb/t)	Mannose (lb/t)	Ethanol (gal/t)
CB pellets	32	194.2	377	39	32.1
CB pellets	33	197.6	423	30	35.0
CB pellets	34	199.8	449	22	36.4
CB pellets	35	201.8	388	17	31.3
Newsprint	32	189.9	204	57	20.2
Newsprint	34	198.7	331	37	28.4
Newsprint	35	200.4	337	25	28.0
Newsprint	36	202.4	282	16	23.0
NP pellets	31	185.4	270	38	23.8
NP pellets	32	189.3	271	33	23.5
NP pellets	33	190.1	232	15	19.1
NP pellets	34	186.2	230	14	18.9
RDF pellets	34	196.8	191	52	18.8
RDF pellets	35	197.9	202	41	18.8
RDF pellets	36	198.9	214	36	19.3
RDF pellets	39	200.2	227	22	19.2

[a] Yields assume 100% sugar recovery, theoretical conversion efficiency and 100% ethanol recovery.
[b] CB = Cardboard, Newsprint = Shredded Newsprint, NP = Newsprint, and RDF = Refuse Derived Fuel
[c] Time from start of test

the maximum temperature. Pressures over 20 atmospheric barrs exceed the pressure limits of the vessel.

These tests were performed using 2% (w/w) sulfuric acid concentration, a 5:1 liquid-to-solid ratio, 1,500 g of dry material, peak temperatures ranging from 185-205° C, and a retention time of up to 45 minutes. Retention times were determined from time of initial start until the end of the test. Previous laboratory tests determined the retention time starting when the material reached the desired temperature until the end of the test and did not include the start-up period. Samples were taken every minute starting at 150° C until the end of the test to determine the maximum sugar production.

From the test results shown in Table II, yields of 36, 28, 24, and 19 gallons of ethanol per dry ton were achieved from cardboard pellets, shredded newsprint, newsprint pellets, and RDF pellets, respectively. The yield for cardboard pellets is higher than the other feedstocks. Feedstock analysis has shown potential glucose content to be higher in cardboard than in the other feedstocks. This may be due to the fact that as cardboard is produced, it is processed more than the other feedstocks which removes more hemicellulose and leaves a greater proportion of cellulose. Also, starch is used in the processing of cardboard which could increase the amount of glucose recovered in the hydrolyzate.

Table II also shows the amount of glucose and mannose produced during the high temperature hydrolysis tests which is the amount of total fermentable sugar (TFS). Only glucose and mannose are included in TFS because they are the sugars fermented to ethanol using traditional yeasts. For the cardboard pellets, maximum TFS occurred at 199.8° C and 34 minutes after initial start-up. The intended temperature of 205° C for this test was reached after 37 minutes and maintained through the end of the test. The maximum TFS production for shredded newsprint occurred at 198.7° C and 34 minutes after initial start-up (the highest temperature reached during this test was 205° C at 44 minutes). For the newsprint pellets, maximum TFS production was at 185.4° C and 31 minutes after initial start-up. The intended temperature for this test was 200° C, but due to an increase in pressure in the digester, 197° C was the highest temperature attained. Maximum TFS production for RDF pellets occurred at 198.9° C and 36 minutes after initial start-up. The target temperature for this test was 200° C. This was reached after 37 minutes and maintained until the end of the test. With the exception of the newsprint pellet tests at minimum, the maximum TFS production for the feedstocks occurred from 34-36 minutes after start-up and at approximately 199° C. The maximum TFS production for the newsprint pellets occurred at a shorter retention time and lower temperature.

On comparing the high temperature hydrolysis with previous hydrolysis tests, there are several important differences to be found. For newsprint, previous hydrolysis tests at 160° C produced yields of 24 gal/ton of ethanol, while high temperature hydrolysis tests produced 28 gal/ton. The newsprint pellet high temperature test produced 5 gal/ton more ethanol than any of the other previous hydrolysis tests. For RDF pellets, previous hydrolysis tests at 180° C produced similar yields. The cardboard and newsprint pellets had not previously been hydrolyzed. These results confirm the fact that, generally, the higher the temperature (up to 200° C) the greater the glucose yield, particularly at short retention times. Both feedstocks produced good yields during the high temperature tests. The RDF pellets from Humboldt are of poor quality, and tests with higher quality RDF are planned. In the demonstration-scale facility, the front-end processing would leave cardboard and newsprint in the feedstock going to hydrolysis (Humboldt does not) which would produce a higher quality feedstock and result in higher RDF-to-ethanol yields. Future tests also may include different acid concentrations and liquid-to-solid ratios (*6*).

Pilot-Plant Results. Previous work by Watson, et al., 1990, (*4*), Broder, et al., 1991, (*7*), and Broder et al., 1991, (*5*) discussed results of TVA pilot-plant tests using waste cellulosics as feedstock. To verify laboratory yields, tests have been conducted on a much larger scale in TVA's 2-ton-per-day pilot plant using RDF pellets, cardboard pellets, and newsprint pellets. These tests were designed to compare the effects of retention time and operating temperature on the yield of fermentable sugars from these feedstocks. Results are shown in Table III. The best yields using the RDF pellet occurred at an operating temperature of 180° C for 15 minutes. These conditions resulted in yields of up to 16 gallons per dry ton of RDF. The cardboard pellets provided yields of up to 29 gallons, and the operating conditions were 160° C for 20 minutes. The newsprint provided a yield of 23 gallons per ton obtained at 160° C and a 10-minute retention time.

The RDF pellets obtained from Humboldt are variable in quality (non-homogeneous). Several of the crates contained significant amounts of unsorted metal, probably aluminum cans. Pieces of metal passed through the feed system without being noticed and caused erratic operation of the discharge valves. This behavior caused greater than normal fluctuations in hydrolyzer temperature and feedstock level in the hydrolyzer. Tests conducted in October and November 1990 with cardboard and newsprint proved they were more homogeneous feedstocks, and the hydrolyzer performed well. After completion of tests with newsprint in February and March 1991, it was found that the screw feed system was working incorrectly, and temperature and retention times were not accurate. This is being corrected, and yields are predicted to be higher in successive runs (*5*).

Fermentation Tests. Laboratory fermentation tests were conducted with newsprint and two RDF hydrolyzates. Table IV contains concentrations of potential fermentation inhibitors, primarily sugar degradation products, found in the substrates. Table V contains results of fermentation tests with the three hydrolyzates. Three treatments were used for substrate preparation to improve fermentation in the presence of these inhibitors. The three treatments were (1) adjustment of pH to a desirable range for yeast fermentation, (2) over-liming prior to pH adjustment (overliming the hydrolyzate to pH 10 and readjusting with sulfuric acid to 5.5), and (3) over-liming plus addition of 1 g/L sodium sulfite. Nutrients known to enhance fermentation were added to the hydrolyzates (2 g/L urea, 2 g/L yeast extract, 0.3 g/L KCL, and 0.4 g/L reagent phosphoric acid). <u>Saccharomyces cerevisiae</u> and <u>Candida utilis</u> were grown in 20 g/L glucose plus 6.7 g/L yeast nitrogen base (Difco) for 24 hours at 30° and 100 rpm on a rotary shaker. Yeasts were recovered by centrifugation and suspended in pH 7 phosphate buffer. Test hydrolyzate preparations were given a 5% by volume inoculum of the respective yeast strains.

Table III. Pilot-plant test results

Nominal Reactor Conditions Temp(C)	Time(min)	Ethanol Yield From Glucose (gal/t)	Total Ethanol Yield (gal/t)
Humboldt RDF Pellets-9/90			
180	15	14.00	16.10
160	20	11.20	12.88
Cardboard Pellets-10/90			
161	20	19.28	29.00
161	25	16.44	25.00
174	20	16.59	23.00
Newsprint Pellets-11/90			
160	10	17.76	23.00
159	20	10.16	16.00
169	15	9.90	17.00
Newsprint Pellets-2/91			
179	10	11.24	16.00
180	20	16.92	20.00
Newsprint Pellets-3/91			
170	15	8.10	14.00
180	10	10.89	16.00

Table IV. Sugar degradation products [a]

	Formic Acid	Acetic Acid	Levulinic Acid	HMF[c]	Furfural
Newsprint	2.9	2.1	6.2	1.3	1.5
RDF[b] #1	4.8	1.4	10.6	1.7	1.6
RDF #2	0.9	1.1	1.5	2.2	1.4

[a] Sugar degradation products given in g/L in liquid fraction.
[b] RDF = Refuse derived fuel
[c] HMF = 5-Hydroxymethyl-2-furfural

The initial growth rates of the yeasts demonstrated a lag for about the first 12 hours. Samples were taken for 8 days due to the slow yeast growth and consequently slow ethanol production rate. The newsprint hydrolyzate showed extreme effects from the treatments, with treatment three resulting in the best yields. The yield of ethanol from glucose and mannose in the two RDF hydrolyzates using Saccharomyces cerevisiae was unaffected by the treatment applied. The ethanol produced accounted for conversion of all the glucose and mannose present.

Candida utilis was more sensitive to the hydrolyzate source and treatment. Fermentation of RDF #1 showed good results only with treatment 3. Fermentation of RDF #2 showed good results with treatments 2 and 3. This reflects the difference in the composition of the hydrolyzates. RDF #1 contained higher levels of levulinic and formic acids than RDF #2, and these are known inhibitors of fermentation by yeasts.

The rate of fermentation was much less than desired throughout all tests. All hydrolyzates receiving treatment 3 were fermented by Saccharomyces cerevisiae in three days.

The sugar concentrations of the hydrolyzates were relatively low. It is desirable to ferment substrates of greater concentration in order to provide a greater product concentration. These tests were also performed using only two yeasts. There may be other species or treatments that will allow a faster rate of ethanol production and still result in the near theoretical yields achieved in these tests.

Safety and Environmental Aspects

Previous work by Barrier, J.W., 1990, (*8*), Broder et al., 1991, (*7*), and Barrier, J.W., 1991, (*9*) discussed results of TVA's dilute sulfuric acid hydrolysis effluent streams. TVA's dilute sulfuric acid hydrolysis process results in three main effluent streams: gypsum, lignin residue, and stillage. Waste streams are considered hazardous if they exhibit any of the Resource Conservation and Recovery Act (RCRA) characteristics which are set forth by the Environmental Protection Agency (EPA). The characteristics are corrosivity, ignitability, reactivity, and leachability. Analyses of these waste streams were conducted by TVA's analytical laboratory. The leachability analyses were done by the EP Toxicity method which has now been replaced by the Toxicity Characteristic Leaching Procedure (TCLP). The TCLP includes 25 additional chemicals (mostly herbicides and pesticides) not previously included.

Gypsum. Analyses required for a solid include the RCRA characteristic tests for corrosivity, leaching, and reactivity. Tables VI and VII show leaching and reactivity characteristics of gypsum. The gypsum was produced from neutralization of RDF hydrolyzate in the laboratory. Corrosivity of a solid is determined by measuring pH, and the pH must be above 2.0 and below 12.5. For gypsum, the measured pH was 4.9-7.6. The analyses shown in Tables VI and VII indicate that gypsum can be safely landfilled since it does not exhibit the RCRA characteristics of a hazardous waste.

Table V. Results of fermentation tests with three hydrolyzates

Substrate[a]	EtOH (g/L)	Best Conc Reached(day)	Sugars Fed (g/L)	Glu+Man (% of total sugars)	Total Sugars Used(%)
TREATMENT 1					
Newsprint-S	0.3	8	35.5	79	3
Newsprint-C	0.0	8	35.5	79	0
RDF[b] #1-S	11.2	8	25.1	86	86
RDF #1-C	0.2	8	25.1	86	16
RDF #2-S	4.6	8	15.9	66	69
RDF #2-C	0.3	8	15.9	69	3
TREATMENT 2					
Newsprint-S	0.2	8	35.3	79	4
Newsprint-C	0.0	8	35.3	79	0
RDF #1-S	9.6	8	23.5	85	86
RDF #1-C	0.5	8	23.5	86	8
RDF #2-S	5.2	8	15.9	68	69
RDF #2-C	5.3	8	15.9	70	68
TREATMENT 3					
Newsprint-S	13.1	3	35.2	79	75
Newsprint-C	7.8	8	35.2	79	45
RDF #1-S	9.5	3	24.8	85	86
RDF #1-C	10.0	8	24.8	85	86
RDF #2-S	4.6	3	16.1	67	68
RDF #2-C	5.4	2	16.1	70	70

[a] The S and C represent Saccharomyces cerevisiae and Candida utilis, respectively.
[b] RDF = Refuse Derived Fuel

Table VI. Leaching characteristics of gypsum and lignin residue using the EP Toxicity analyses

Chemical	Gypsum (ppm)		Residue (ppm)		Maximum Limit (ppm)
	S1	S2	S3	S4	
Arsenic	<0.23	<0.23	<0.23	<0.23	5.0
Barium	0.029	0.016	0.463	0.275	100.0
Cadmium	<0.024	<0.024	<0.024	<0.024	1.0
Chromium	2.35	0.166	<0.016	<0.016	5.0
Lead	<0.061	<0.061	<0.061	<0.061	5.0
Mercury	<0.052	<0.052	<0.052	<0.052	0.2
Selenium	<0.256	<0.256	<0.256	<0.256	1.0
Silver	<0.007	<0.007	<0.007	<0.007	5.0
Endrin	<0.005	<0.005	<0.005	<0.005	0.02
Lindane	<0.001	<0.001	<0.001	<0.001	0.4
Methoxychlor	<0.05	<0.05	<0.05	<0.05	10.0
Toxaphene	<0.25	<0.25	<0.25	<0.25	0.5
2,4-D	<0.40	<0.40	<0.40	<0.40	10.0
2,4,5-TP	<0.40	<0.40	<0.40	<0.40	1.0

Note: S1 and S2 are two samples where S1 has been neutralized to a pH of 5.5 and S2 has been over-limed to a pH of 10.0. S3 and S4 are similar samples except that sample S3 has had inerts removed.

Lignin Residue. The lignin residue produced from hydrolysis of RDF in the laboratory was analyzed. To recover maximum soluble sugars, lignin residue is washed, and the resulting pH was 2.5- 4.0. Results of the leaching and reactivity tests are shown in Tables VI and VII. These results indicate that the lignin residue does not exhibit any of the RCRA characteristics of hazardous wastes, and the residue can therefore be considered a solid waste.

Stillage. Tables VII-IX show analyses of the stillage. Stillage contains from 96-99% water. Hydrolyzate was analyzed as being representative of stillage. Hydrolyzate is easy to produce in large volumes for analytical purposes and becomes stillage after fermentation and removal of ethanol. As a liquid, the hydrolyzate was analyzed for the RCRA characteristics of reactivity, corrosivity, and ignitability. The analyses of the hydrolyzate indicate that it does not exhibit any of the RCRA characteristics of a hazardous waste.

The analyses of TVA's dilute sulfuric acid hydrolysis processing system have shown the production of no hazardous waste streams. However, since the composition of MSW and RDF is variable, the process effluent streams will continue to be analyzed on a regular basis to ensure safety and environmental acceptability (7).

Economics

Previous work by Bulls et al., 1991, (*10*), Barrier, J.W., 1990, (*11*), Broder et al., 1991 (*7*), Barrier, J.W., 1991, (*9*), Barrier et al., 1991, (*12*), Barrier, J.W., 1991, (*13*), discussed results of TVA's technical economic evaluations of its MSW processing system. An estimate of total capital investment for the MSW processing system for a 1,000-ton-a-day processing system is shown in Table X. These capital costs are based on vendor quotes for the major equipment items and include direct and indirect costs such as piping, electrical, engineering and supervision, contingency, etc. A total capital investment of $104 million has been estimated for the plant design. Operating costs for the plant are shown in Table XI. Raw material costs include sulfuric acid, lime, yeast, nutrients for fermentation, and an ethanol denaturant. Utilities, landfilling, labor, supplies, and fixed charges make up the balance of the operating costs. Fixed charges include depreciation (straight line, 20 years), insurance, taxes, maintenance, and plant overhead. Because of the tipping fee associated with the disposal of MSW, the "cost" for this feedstock is negative and provides a credit to the process.

Revenue from the process is based on the sale of recyclables, chemicals, and electricity. Recyclables include aluminum, ferrous metals, plastics, and glass. Price estimates are based on national averages for these items. As shown in Table XI, total annual revenue from recyclables is estimated to be $7.8 million. Chemicals produced in the process include ethanol, furfural, acetic acid, and carbon dioxide. Prices for these items were obtained from industry quotes and the <u>Chemical Marketing Reporter</u>. Total revenue from chemicals produced in the process is estimated to be $8.5 million per year. Annual revenue from excess electricity is estimated to be $4.7 million, based on $0.06 per kWh.

Table VII. Reactivity analyses of gypsum, lignin residue, and hydrolyzate for cyanide and sulfide production

Sample	Reactivity(ppm)		Maximum Limits(ppm)	
	CN-	S-	CN-	S-
Gypsum	< 0.05	< < 500	250	500
Lignin Residue	< 0.05	< < 500	250	500
Hydrolyzate	< 0.05	< 500	250	500

Table VIII. Corrosivity analyses of neutralized hydrolyzate for corroding steel (SAE 1020)

Sample	Corrosivity (inch/yr)	Maximum Limit (inch/yr)
Hydrolyzate 1	0.00895	0.250
Hydrolyzate 2	0.01754	0.250

Table IX. Ignitability of neutralized hydrolyzate

Sample	Ignitability Flash Point	Minimum Limit
Hydrolyzate 1	> 60° C	60° C
Hydrolyzate 2	> 60° C	60° C

Table X. Total capital investment for a 1,000-ton-per-day MSW processing facility

ITEM	COST ($)
Direct Costs	
Purchased Equipment-Delivered	5,784,355
Equipment Installation	7,735,306
Instrumentation/Controls/Electrical/Piping	9,798,054
Buildings and Yard Improvements	7,219,620
Service Facilities	12,118,647
Land	1,547,061
Total Direct Costs	64,203,044
Indirect Costs	
Engineering and Supervision	5,930,402
Construction Expenses	6,446,089
Total Indirect Costs	12,376,491
Contractor's Fee	3,094,123
Contingency	7,735,306
Working Capital	16,759,831
TOTAL CAPITAL INVESTMENT	104,168,794

Table XI. Production costs for TVA's ethanol-from-MSW process for a 1,000-ton-per-day plant (quantities are in tons unless otherwise noted)

ITEM	QUANTITY	$/UNIT	AMOUNT($/yr)
COSTS			
MSW	330,000	-38.37	(12,662,065)
Sulfuric Acid	8,865	70.00	620,537
Lime	6,493	45.00	292,186
Yeast	414	150.00	62,039
Potas.Dihy.Phos.	208	50.00	10,423
Urea	827	115.00	95,127
Sodium Sulfite	414	475.00	196,458
Denaturant-GAL	221,528	0.50	110,764
Utilities			
-Process Water-MG	40	500.0	020,001
-Cooling Water-MG	615	50.00	30,771
Landfilling	87,737	30.00	2,632,113
Labor	---	---	2,640,538
Supplies	---	---	524,454
Fixed Charges			
-Depreciation-20 YR	---	---	4,293,095
-Insurance	---	---	874,090
-Local Taxes	---	---	874,090
-Maintenance	---	---	3,496,359
-Plant Overhead	---	---	1,320,269
Total Costs			5,431,249
REVENUES			
Recyclables			
-Aluminum	4,158	1,000.00	4,158,000
-Glass	30,490	15.00	457,352
-HDPE	8,465	140.00	1,185,030
-PET	4,389	140.00	614,460
-Ferrous Metals	27,786	50.00	1,389,300
Ethanol-GAL	4,652,086	1.25	5,815,107
Carbon Dioxide	14,592	10.00	145,916
Furfural	2,364	800.00	1,891,568
Acetic Acid	1,107	580.00	641,865
Electricity			
-KWH/YR	7.93×10^7	0.06	4,757,969
Total Revenues			21,056,568
RATE OF RETURN ON INVESTMENT			15.00%

Based on the costs and revenues associated with the process, and a required return on investment of 15%, the required tipping fee was calculated to be $38.37/ton of MSW (10).

Conclusions

Hydrolysis conditions will influence sugar production, fermentation efficiency, and effluent composition. Laboratory tests have shown yields of up to 27 gallons of ethanol can be produced per ton of waste-derived feedstock and 36 gallons per ton of cardboard pellet feedstock. Pilot-plant tests have shown yields up to 16 gallons of ethanol per ton of waste-derived feedstock. Further tests at higher acid concentration, lower liquid-to-solid ratios, higher temperatures and shorter retention times may increase sugar yields, produce more concentrated hydrolyzates and reduce the quantity of residue remaining after hydrolysis. Significant quantities of organic residue remain after hydrolysis which are being evaluated as an energy source to run the process.

Municipal solid waste varies in composition, and analyses will be performed on a regular basis on each supply of MSW. A replicated study is planned to evaluate variability and component distribution in the different processing streams. Analyses of processing effluents from the conversion of RDF in TVA's process has resulted in the production of no hazardous materials based on Environmental Protection Agency (EPA) guidelines. Work is in progress to characterize gases, volatile organics, and semivolatile organics from hydrolysis, fermentation, and residue combustion. Many questions related to process scale-up and effluent management remain to be answered.

Preliminary economic analysis shows a tipping fee of $38.37 a ton is required for MSW conversion using a 15% rate of return on investment. Additional economic evaluations will be conducted as the process is optimized.

Literature Cited

1. National Solid Waste Management Association. "*Solid Waste Disposal Overview*." 1989.
2. National Solid Waste Management Association. "*Landfill Capacity in the Year 2000*." 1989.
3. Barrier, J.W., Bulls, M.M., Broder, J.D., Lambert, R.O., "*Production of Ethanol and Coproducts from MSW-Derived Cellulosics Using Dilute Sulfuric Acid Hydrolysis*." Presented at the Twelfth Symposium on Biotechnology for Fuels and Chemicals, Gatlinburg, Tennessee, May 1990.
4. Strickland, R.C., Watson, J.R., "*Summary of Ethanol-from-MSW Yield Data for the Second Quarter of the Project*." July-September 1990, TVA Quarterly Progress Report.
5. Broder, J.D., Strickland, R.C., Lambert, R.O., Bulls, M.M., Barrier, J.W. "*Biofuels from Municipal Solid Waste*." ASME Mtg, San Diego, California, October 1991.

6. Tuten, D.S., Softley, L.G., Strickland, R.C. "*High Temperature Hydrolysis Tests For Newsprint Pellets in the Rotating Digester.*" April-June 1991, Quarterly Progress Report.

7. Broder, J.D. Barrier, J.W., Bulls, M.M. "*Producing Fuel Ethanol and Other Chemicals from Municipal Solid Wastes.*" Presented at The American Society of Agricultural Engineers Summer International Meeting, Albuquerque, New Mexico, June 1991.

8. Barrier, J.W., "*Municipal Solid Waste and Waste Cellulosics Conversion to Fuels and Chemicals.*" July-September 1990, TVA Quarterly Progress Report.

9. Barrier, J.W., "*Municipal Solid Waste and Waste Cellulosics Conversion to Fuels and Chemicals.*" April-June 1991, TVA Quarterly Progress Report.

10. Bulls, M.M., Shipley, T.M., Barrier, J.W., Lambert, R.O., Broder, J.D. "*Comparison of MSW Utilization Technologies--Ethanol Production, RDF Combustion, and Mass Burning.*" Presented at the Southern Biomass Conference in Baton Rouge, Louisiana, January 1991.

11. Barrier, J.W., "*Municipal Solid Waste and Waste Cellulosics Conversion to Fuels and Chemicals.*" October-December 1990, TVA Quarterly Progress Report.

12. Barrier, J.W., Bulls, M.M., Shipley, T.M. "*Ethanol Produc tion from Refuse Derived Waste Using Dilute Sulfuric Acid Hydrolysis.*" Presented at the National Bioenergy Conference in Coeur d'Alene, Idaho, March 1991.

13. Barrier, J.W., "*Municipal Solid Waste and Waste Cellulosics Conversion to Fuels and Chemicals.*" January-March 1991, TVA Quarterly Progress Report.

RECEIVED June 15, 1992

Chapter 5

Fuel Evaluation for a Fluidized-Bed Gasification Process (U-GAS)

A. Goyal and A. Rehmat

**Institute of Gas Technology, 3424 South State Street
Chicago, IL 60616**

The gasification characteristics of a solid carbonaceous fuel in the U-GAS fluidized-bed gasification process can be predicted by laboratory examination of the fuel, which includes chemical and physical characterization and thermobalance and agglomeration bench-scale characterization and bench-scale tests. Additional design information can be obtained by testing the feedstock in the U-GAS process development unit or the pilot plant.

The Institute of Gas Technology (IGT) has developed an advanced, single-stage, fluidized-bed gasification process, the U-GAS process, to produce a low- to medium-Btu gas from a variety of solid carbonaceous feedstocks, such as coal, peat, wood/biomass, sludge, etc. The development of the process is based on extensive laboratory testing of these feedstocks as well as large-scale tests in a low-pressure (50 psig) pilot plant and a high-pressure (450 psig) process development unit conducted over a period of several years. Up to 98% feedstock utilization with long-term steady-state operation has been achieved. The testing has provided information related to the effect of various gasification parameters, such as pressure, temperature, and steam-to-carbon feed ratio, on gasification characteristics of the feedstocks. The concept of *in-situ* desulfurization by simultaneous feeding of dolomite/limestone has also been established. Reliable techniques have been developed for start-up, shutdown, turndown, and process control. The process represents the fruition of research and development in progress at IGT since 1974. The product gas will be a low-Btu gas that is usable as a fuel when operating with air, and a medium-Btu or synthesis gas when operating with oxygen. The medium-Btu or synthesis gas can be used directly as a fuel, converted to substitute natural gas, or used for the production of chemical products such as ammonia, methanol, hydrogen, and oxo-chemicals. The low- and medium-Btu gas can also be used to produce electricity generated by a combined cycle or by fuel cells.

0097–6156/93/0515–0058$06.00/0

On the basis of the operational results with numerous feedstocks, IGT has developed an experimental program for the evaluation of a solid carbonaceous fuel for use in its fluidized-bed gasification technology.

U-GAS Process

The U-GAS process employs an advanced, single-stage, fluidized-bed gasifier (Figure 1). The feedstock, which is dried only to the extent required for handling purposes, is pneumatically injected into the gasifier through a lockhopper system. Within the fluidized bed, the feedstock reacts with steam and air or oxygen at a temperature dictated by the feedstock characteristics; the temperature is controlled to maintain nonslagging conditions of ash. The gases are introduced into the gasifier at different compositions at different points at the bottom of the gasifier. The operating pressure of the process depends on the ultimate use of the product gas and may vary between 50 and 450 psi. Upon introduction, the feedstock is gasified rapidly and produces a gas mixture of hydrogen, carbon monoxide, carbon dioxide, water, and methane, in addition to hydrogen sulfide and other trace impurities. Because reducing conditions are maintained in the bed, nearly all of the sulfur present in the feedstock is converted to hydrogen sulfide.

The fines elutriated from the fluidized bed are separated from the product gas in two stages of external cyclones and are returned to the bed where they are gasified to extinction. The product gas is virtually free of tars and oils due to the relatively high temperature of the fluidized-bed operation, which simplifies the ensuing heat recovery and gas cleanup steps. The process yields a high conversion, especially because of its ability to produce ash agglomerates from some of the feedstocks and selective discharge of these agglomerates from the fluidized bed of char.

Fuel Evaluation

Three steps are recommended to evaluate the suitability of a potential feedstock for the process:

1. Laboratory analyses

2. Bench-scale tests

3. Process development unit (PDU) or pilot plant gasification test.

Laboratory Analyses. Table I lists those fuel properties that are normally determined for assessing a solid fuel for use in the process. Additional analyses are performed as required with unusual feedstocks. For example, run-of-mine coals with a high mineral content may require mineral identification and evaluation of the effect of high mineral content on the ash fusion properties.

The bulk density, heating value, ash content, and elemental composition of the organic portion of the feedstock usually have no direct effect on the behavior

Table I. Laboratory Analyses of the Fuel

- Proximate Analysis
- Ultimate Analysis
- Higher/Lower Heating Value
- Bulk Density
- Particle-Size Distribution
- Grindability
- Equilibrium Moisture
- Free Swelling Index
- Ash Fusion Temperatures (Reducing Atmosphere)
- Ash Mineral Analysis

of the feedstock in fluidized-bed gasification. However, they do influence the oxygen requirement, the gas yield, and the gas composition. The higher heating value (HHV) is a measure of the energy content of the feedstock. It relates, with other factors, to the amount of oxygen needed to provide the desired gasification temperature levels. If a feedstock has a low HHV, more oxygen is needed to maintain the gasifier temperature at an acceptable operating level. If the HHV is higher, less oxygen will be required to maintain the desired temperature levels.

The ash fusion temperature reflects the ease of agglomeration of the ash in the gasifier. The free swelling index (FSI) indicates the caking tendency of the feedstock; for highly caking feedstocks, a proper distribution of the feed material, as it enters the gasifier, is critical. In the U-GAS process, the Pittsburgh No. 8 bituminous coal with an FSI of 8 has been successfully gasified and agglomerated with overall coal utilization of 96%. The feedstock is generally sized to 1/4-inch X 0 before it is fed to the gasifier. If a finer size is available, the fluidization velocity is reduced accordingly.

To utilize a feedstock today, one needs to know a great deal about it prior to purchase. It is essential to know the sulfur content to comply with airborne emissions standards and the ash content and its constituents to ensure compliance with solid waste regulations. Other standards are still evolving as new environmental and energy legislation is enacted.

The range of various properties of the feedstocks that have been tested in the U-GAS process development unit or pilot plant is given in Table II.

Bench-Scale Tests. Three types of bench-scale tests are conducted to evaluate the fuel. These bench-scale tests establish a range of operating conditions that can be used to plan tests in the process development unit or the pilot plant facility, and to perform material and energy balances for the gasifier and estimate its throughput. These tests are described below.

Thermobalance Tests. The gasification of a solid carbonaceous fuel consists of two major steps: 1) initial rapid pyrolysis of the feedstock to produce char, gases, and tar and 2) the subsequent gasification of the char produced. (In addition, some combustion reactions take place if gaseous oxygen is present; these reactions

**Table II. Range of Feedstock Properties
Tested in the U-GAS Process**

Moisture,* %	0.2 to 41
Volatile Matter,** %	3 to 69
Ash,** %	6 to 78
Sulfur,** %	0.2 to 4.6
Free Swelling Index	0 to 8
Ash Softening Temperature, °F	1980 to 2490
Higher Heating Value,** Btu/lb	2,330 to 13,630

*   As received.
** Dry basis.

are very rapid.) Because the rate of the second step is much slower than that of the first step, the volume of a gasifier (or the carbon conversion in the gasifier) is primarily dependent on the gasification rate of the char. Due to the relatively well-mixed nature of a fluidized-bed gasifier, the char particles undergoing gasification are exposed to gases consisting primarily of CO, CO_2, H_2, H_2O, and N_2.

The thermobalance testing is performed to determine a relative reactivity constant for the feedstock for comparison with the reference coal, Western Kentucky No. 9 bituminous coal, which has been extensively tested in the thermobalance (*1*) as well as in the U-GAS process. In the thermobalance, a small quantity of the feedstock is continuously weighed while being gasified at a specific temperature, pressure, and gas composition. This measured weight loss data versus time and the thermobalance operating conditions and analyses of feed and residue are used to calculate the specific relative reactivity constant for the feedstock. The kinetic data, in conjunction with the reference coal information, are used to plan tests in the PDU or pilot plant. As an example, Figure 2 shows the gasification rate for maple hardwood char, peat char, and bituminous coal char, as determined by the thermobalance. The carbon conversion data for Western Kentucky No. 9 coal char under different operating conditions, as determined by the thermobalance, are shown in Figure 3.

Ash-Agglomeration Tests. Prior to the large-scale testing, the ash-agglomeration tests are conducted in the laboratory to determine the possibility of agglomerating the feedstock ash in the gasifier. These tests are performed in a 2-inch fluidized-bed reactor capable of operating at temperatures up to 2200°F. Several tests have successfully demonstrated that ash agglomerates can be produced in this bench-scale unit at conditions that can be related to the pilot plant operating conditions. The 2-inch reactor has a unique grid design that allows close

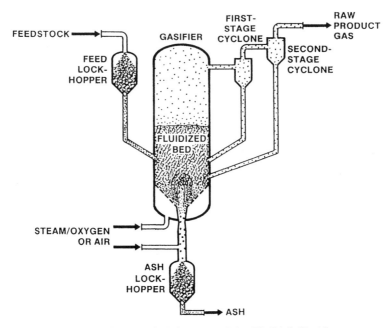

Figure 1. Schematic Diagram of the U-GAS Gasifier

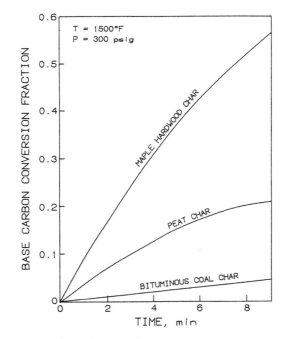

Figure 2. Gasification Rates for Maple Hardwood Char, Peat Char, and Bituminous Coal Char

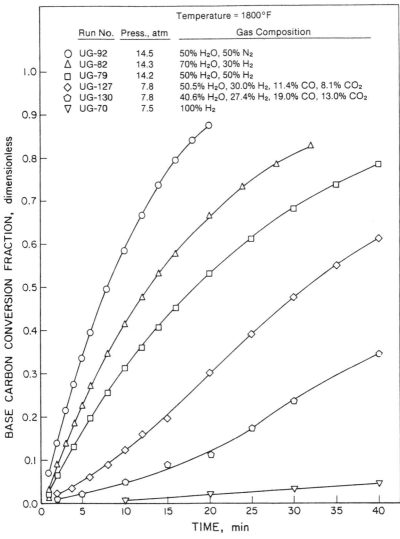

Figure 3. Base Carbon Conversion Fraction Versus Time for Gasification of Western Kentucky No. 9 Coal Char

simulation of the pilot plant fluidized-bed dynamics and mixing characteristics, which are essential for proper ash agglomerate formation, growth, and discharge. The tests are generally conducted at different temperatures, superficial velocities, gas compositions, and operating times to evaluate conditions favoring ash agglomerate formation and growth. The results are quantified using size distribution curves of feed, residue, and fines to show size growth of particles. Visual evaluation of the agglomerates includes separation of the +8 mesh fraction (normally 100% agglomerates) in the residue and, if required, separation of agglomerates by float-sink techniques for each size fraction. The agglomerates thus separated can be easily photographed or examined petrographically. An example of the test results with different coal samples is given in Table III.

Table III. 2-Inch Ash-Agglomeration Tests With Various Feedstocks

Coal Sample	Temp., °F	Char Initial Ash, %	Run Time, h	Fluidizing Velocity, ft/s	Comments
FC-1	1985	31.5	2.0	1.0	Sinter particles plus some agglomerates
FC-1	2100	31.5	1.5	1.5	Agglomerates formed, little or no sinter
FC-2	1990	45.5	1.0	1.5	Small agglomerates present
FC-2	1990	45.5	1.3	1.5	Larger agglomerates found
FC-3	2080	45.4	2.5	1.5	Agglomerates formed
FC-4	1960	15.5	1.0	2.1	No agglomerates found
FC-4	1920	20.9	3.0	1.5	Small agglomerates found
FC-4	2000	15.5	2.5	1.6	Greater number of large agglomerates
KY #9	2000	51.0	1.3	1.5	Many agglomerates produced

Fluidization Test. A fluidization test in a glass column at ambient conditions may also be conducted to determine the minimum and complete fluidization velocities of the material. This information is then translated into the necessary operating velocity in the PDU or pilot plant test. The fluidization test is conducted only if the feedstock is unusual or if the feedstock size is different than that typically used (1/4 in. X 0) in the process. This test is conducted with the char produced from the feedstock.

Process Development Unit (PDU) or Pilot Plant Gasification Test. IGT has two continuous U-GAS gasification units located in Chicago: 1) An 8-inch/12-inch dual-diameter high-pressure process development unit, which can be operated at up to 450 psig and has a nominal capacity of 10 tons per day (at 450 psig operation), and 2) A 3-foot-diameter low-pressure pilot plant, which can be operated at up to 50 psig and has a nominal capacity of 30 tons per day. A process flow schematic of the U-GAS pilot plant is shown in Figure 4. In addition, a 2-foot/3-foot dual-diameter high-pressure pilot plant has recently been constructed at Tampere, Finland, and testing in this unit has begun. Plans are under way to test various coals, peat, wood and bark waste, and pulp mill sludge in this unit.

A test in the PDU or pilot plant provides the following information:

*   It confirms the suitability of the candidate feedstock for the U-GAS process.

*   It establishes the base design operating conditions as well as an operating window for the gasifier.

*   Design data for fines characteristics, ash agglomeration characteristics, and gas characteristics are obtained.

*   Estimates for gas quality, gas yields, and process efficiency are established.

*   Necessary environmental data to define the environmental impact are taken.

*   Various samples, such as bed material samples, ash discharge samples, fines samples, and wastewater samples, are collected and saved and provided as needed for use during detailed engineering.

The PDU testing is recommended where high-pressure gasifier operation would be required. Each test in the PDU usually consists of 2 days of operation, whereas one 5-day-duration test is usually conducted in the pilot plant with the candidate feedstock. During the test, the gasifier is operated in ash-balanced, steady-state conditions, during which most of the design data are procured. A detailed test plan is generally prepared based on a comparison of the feedstock with a similar feedstock or from information obtained from bench-scale testing. Depending on the feedstock characteristics, the gasifier is operated at a temperature of up to 2000°F and a superficial velocity of up to 5 ft/s.

Numerous solids samples are collected regularly during the test run so that accurate material balances can be prepared. Process sample points include the coal feed, fluidized bed, ash discharge, and cyclone diplegs (for the pilot plant). Samples from the fluidized bed are also collected and analyzed hourly during the test to help the operators determine and maintain steady-state operation.

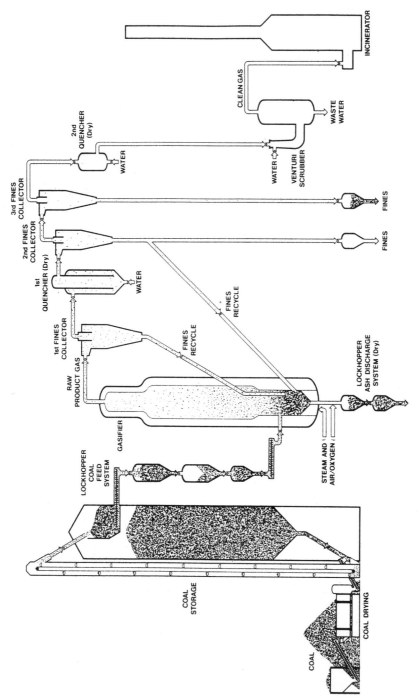

Figure 4. Process Flow Schematic of the U-GAS Pilot Plant

All process solids and gas flow streams are measured and recorded. Temperatures are recorded for all process streams and at several locations within the reactor. Redundancy is provided for the reactor pressure taps used for bed density and height.

A product gas sample stream is drawn continuously from the gasifier freeboard for chromatograph analysis. The chromatograph system provides accurate on-line analysis for CO, CO_2, CH_4, H_2, N_2, H_2O, and H_2S. The chromatograph sequencing is microprocessor-controlled for flexibility in the scope and frequency of the analysis. The product gas samples are also collected in gas bombs for later laboratory analyses.

Special sampling and instrumentation are available for complete chemical characterization (organic compounds as well as trace elements) of the product gas and wastewater streams. Test results of this nature are necessary to satisfy environmental permitting requirements and for proper design of downstream processing equipment. Equipment is also available for determination of product gas dust loading after one, two, or three stages of cyclone separation.

These units use a microprocessor-based data acquisition system to ensure accurate, timely, and reliable collection of all process data of interest. About 85 process data points (temperature, pressure, flow, etc.) for the pilot plant and about 40 data points for the PDU are scanned repeatedly throughout the test. A full scan is completed in approximately 10 seconds and is repeated at 3-minute intervals. The reactor operating status, including various flows, pressures, temperatures, and velocities (grid, venturi, bed, freeboard, cyclone), bed density, bed height, etc., is calculated and displayed on the computer CRT screen. The data are stored on magnetic tape in both raw signal and converted form. The converted data are averaged hourly, and an hourly average report of all data points and the operating status is automatically printed in the control room; in addition, a shift report is printed every 8 hours to allow a shift engineer to review the operation of a previous shift. Particular emphasis is placed on the use of the data acquisition system as an operating tool. Specialized programs have been developed to aid the operators in the approach to and confirmation of steady-state operation. This results in more steady-state operating time, and therefore more useful design data, per test run.

Table IV summarizes various feedstocks tested in the pilot plant and PDU and the results obtained. The details of the PDU system and some test results are given by Goyal *et al.(2,3)*. The details of the pilot plant system and some test results are given by Goyal and Rehmat *(4,5)*.

The lowest rank material gasified so far in the U-GAS process is peat; some details are given here. Kemira OY, a chemical company in Finland, was interested in producing synthesis gas from peat for manufacturing ammonia. Two different types of peat were received by IGT from Kemira: Viidansuo and Savaloneva. The two samples represented the extremes in peat quality, with Viidansuo being of poorer quality, very fibrous, and having a van Post index of only 3 to 4. Both peats were obtained by crushing peat sods to 3/8 in. X 0 size. The peats as received contained about 40 weight percent moisture. The two peats were mixed for the gasification test, one batch containing equal parts of each and dried to 35 weight percent moisture and another batch containing two parts Savaloneva to one part of Viidansuo and dried to 15 weight percent moisture. Despite the different flow properties of these peats, the U-GAS pilot plant lockhopper feed system, after some modifications, was able to feed them at controlled rates for the steady-state duration of the test. The pilot plant was operated for 5-1/2 days from start-up to shutdown. Forty-two tons of the peat were gasified during 50 hours of continuous operation. The test was conducted at three different operating set points at two different temperatures and at a lower steam feed rate. A peat conversion efficiency of 95% to 97% was maintained throughout the three set points. Figure 5 shows a typical balance for the test. After completion of the three set points, the gasifier operating temperature was raised to obtain data on producing a synthesis gas with a lower methane content. The operation of the gasifier throughout the test was stable and easily controlled by adjusting the peat feed rate and oxygen flow to the gasifier. During interruptions in either peat feed or oxygen feed, the gasifier was able to respond in a controlled and logical manner. The recycle of fines entrained from the gasifier was smooth and very effective, as indicated by high peat conversion efficiency (up to 97%). The peat contained a relatively large fraction of discrete particles of quartz stone. The gasifier bottom, because of its simple design, was able to discharge both the stone and peat ash without any problems. The gasifier was operated under stable, ash-balanced conditions despite a wide variation in the operating conditions by discharging the required amounts of ash through the gasifier bottom section. The test showed that synthesis gas can be produced from the two extremes of the peat quality tested. Even the highly fibrous, undercomposed, and reactive Viidansuo peat was gasified without any problem. In addition, the ash properties -- both iron content and fusion temperature -- of the two peats were quite different, indicating that the gasifier can handle peats of a wide range of ash properties. The test also indicated that a minimal amount of peat preparation in the form of drying and crushing is adequate for feed requirements to the U-GAS process. It is not necessary to either pelletize or deep-dry the peat.

Like coal and peat, it is also possible to gasify wood and bark wastes, refuse-derived fuels, sludge, etc. in the U-GAS fluidized-bed gasifier. IGT has extensive experience in the gasification of such materials in its RENUGAS process, which also employs a single-stage fluidized-bed reactor (6-10). The chemistry of converting the waste materials to synthesis gas is quite similar to that of coal and peat gasification. Since there is normally a cost associated with disposal of the wastes, gasification offers an economically attractive application of the technology. Currently, producers of many of the waste materials must pay a premium for their removal. Ultimate disposal of such materials in a gasification facility would prove to be economically attractive.

In September of 1989, IGT entered into a licensing agreement with Tampella Power, Inc. of Tampere, Finland, which will result in the commercial application of the process. Tampella selected the pressurized fluidized-bed technology because of its versatility and applicability to a wide variety of feedstocks, including coal, peat, forestry waste, etc. As a first step toward commercialization, a 10-MW thermal input pressurized (450-psi) pilot plant has been designed and constructed at Tampella's R&D Center in Tampere, Finland, and testing in this unit has begun. In this plant provisions have been made for testing a variety of feedstocks other

Table IV. Summary of U-GAS Pilot Plant and PDU Testing

Feedstock	Hours of Operation	Total Tons Processed	Year	MAF Coal Utilization,* %
A. Pilot Plant				
Coke Breeze and Coal Chars	4700	1100	1974-76	92
Montana Subbituminous "C" Coal	215	170	1976	92
Illinois No. 6 Bituminous Coal	76	40	1977	86
W. Kentucky No. 9 Bituminous Coal	1303	1031	1978-82	98
ROM W. Kentucky No. 9 Bituminous Coal	442	287	1979	87
Pittsburgh No. 8 Bituminous Coal	173	110	1980	96
W. Kentucky Nos. 9 and 11 Mixed Bituminous Coal	58	52	1980	88
Australian Bituminous Coal	62	39	1981	95
Wyoming Subbituminous "A" Coal	77	84	1981	98
Polish Bituminous Coal	74	46	1981	79
French Bituminous Coal	60	48	1983	95
Unwashed Utah Bituminous Coal	120	58	1984	99
Finnish Peat	50	42	1985	97
Total	7410	3107		
B. Process Development Unit (PDU)*				
Pittsburgh No. 8 Bituminous Coal	7	0.5	1985	89
Illinois No. 6 Bituminous Coal	60	4	1986	86
Montana Rosebud Subbituminous Coal	169	31	1985-86	93
North Dakota Lignite	84	14	1985-86	91
Pittsburgh Coal (fed with limestone)	95	9	1987	97
Indiana New Albany Oil Shale	29	6	1987	89
Total	444	64.5		

* The fines escaping the gasifier are not recycled to the gasifier in the PDU. The coal utilization reported includes the char loss in these fines.

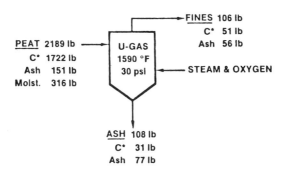

Figure 5. Typical Balance for U-GAS Pilot Plant Test with Finnish Peat. C* represents moisture- and ash-free peat (MAF). Peat conversion = 95.2%.

than coal, including wood and bark waste, peat, and pulpmill sludge. These fuels can be tested separately or as multifuel mixtures. The plant is designed to operate in several gasification modes, including air-blown, enriched air, and oxygen-blown gasification. Provisions have also been made for in-bed desulfurization testing with sorbents such as limestone or dolomite fed directly to the gasifier with the fuel. The pilot plant also provides a versatile testing platform for development and demonstration of various hot-gas cleanup strategies. Included in the plant are high-temperature cyclones for fines removal and recirculation to the gasifier and a high-temperature ceramic filter for complete removal of any remaining fines. A post-bed desulfurization system can also be tested in the plant, allowing testing of the process with both high- and low-sulfur fuels under a wide variety of operating conditions. All fuel gas produced in the plant will be combusted in a waste heat boiler at the end of the gas cleanup train. In addition to providing steam to the process and heat to the city of Tampere via its district heating system, this boiler will allow the combustion and environmental characterization of the fuel gas and combustion products.

Literature Cited

1.   Goyal, A.; Zabransky, R. F.; and Rehmat, A. "Gasification Kinetics of Western Kentucky Bituminous Coal Char," Ind. Eng. Chem. Res. 28, No. 12, pp. 1767-1778 (1989) December.

2.   Goyal, A.; Bryan, B.; and Rehmat, A. "Gasification of a Low-Rank Coal," in Proceedings of the Fifteenth Biennial Low-Rank Fuels Symposium, DOE/METC-90/6109, CONF-890573, pp. 447-464, May 1989.

3.   Goyal, A.; Bryan, B.; and Rehmat, A. "High-Pressure Gasification of Mon tana Subbituminous Coal," in Proceedings of the Sixteenth Biennial Low-Rank Fuels Symposium, LRF 16-91, pp. 302-320, May 1991.

4.   Goyal, A.; Rehmat, A. "The U-GAS Process -- From Research to Commercialization," Paper presented at the AIChE Annual Meeting, San Francisco, California, November 25-30, 1984.

5.   Goyal, A.; Rehmat, A. "Recent Advances in the U-GAS Process," Paper presented at the 1985 Summer National AIChE Meeting, Seattle, Washington, August 25-28, 1985.

6.   Babu, S.; Anderson, G. L.; and Nandi S. P. "Process for Gasification of Cellulosic Biomass," U.S. Patent No. 4,592,762, June 3, 1986, and U.S. Patent No. 4,699,632, October 13, 1987.

7.   Evans, R. J.; Knight, R. A.; Onischak, M.; and Babu S. P. "Development of Biomass Gasification to Produce Substitute Fuels," Final Report to U.S. DOE, Biofuel and Municipal Waste Technology Division, Contract No. B-21-A-O, March 1987.

8.   Onischak, M.; Knight, R. A.; Evans, R. J.; and Babu, S. P. "Gasification of RDF in a Pressurized Fluidized Bed," in Proceedings of Energy From Biomass and Wastes XI, pp. 531-548, March 1987.

9. Onischak, M.; Lynch, P. A.; and Babu, S. P. "Process and Economic Considerations for the Gasification of Pulp Mill Wastes Using the RENUGAS Process," in <u>Proceedings of Energy From Biomass and Wastes XII</u>, pp. 711-730, February 1988.

10. Trenka, A. R.; Kinoshita, C. M.; Takahashi, P. K.; Phillips, V. D.; Caldwell, C.; Kwok, R.; Onischak, M.; and Babu, S. P. "Demonstration Plant for Pressurized Gasification of Biomass Feedstocks," in <u>Proceedings of Energy From Biomass and Wastes XV</u>, pp. 1051-1061, March 1991.

RECEIVED October 14, 1992

Chapter 6

Treatment of Municipal Solid Waste by the HYDROCARB Process

Meyer Steinberg

Process Sciences Division, Department of Applied Science,
Brookhaven National Laboratory, Upton, NY 11973

The HYDROCARB Process addresses the problem of converting
municipal solid waste (MSW) to a clean useable fuel for boilers
and heat engines. The process consists of hydrogenation of the biomass to produce methane, followed by decomposition of the methane to carbon and hydrogen and combining the CO and H_2 to
produce methanol and recycling the hydrogen-rich process gas. Using natural gas to make up the mass balance, the economics looks
attractive for a 3,000 T/D MSW plant, especially when avoidance
costs are taken into account. The process is environmentally attractive since processing is performed in a highly reducing atmosphere
and at elevated pressure and temperatures where no toxic gases are
expected and CO_2 emission is minimized.

It is now generally known that the municipal solid waste problem has become an
ever increasing problem in populated areas in the U.S. The increase in the standard of living manifested by a vast array of consumer goods has added to the
problem of disposal of industrial and municipal solid waste (MSW). The land-fill
disposal sites around metropolitan areas have become less available so that tipping fees are soaring. Municipalities are opting for more waste incineration or
mass-burn plants. Legislation is being passed to require separation of waste for
recycling and resource recovery. Because separated recyclable material is market
demand dependent, the cost of recycling is site specific and time dependent. In
fact, there are a number of municipalities that pay carters to remove and transport recyclable material to other locations which instead of becoming a source of
income becomes a liability.

MSW roughly consists of 50% paper and plastic and the remainder being
glass, metal and kitchen and yard waste. Industrial waste includes paper, wood,
metal and used rubber tires etc. The most traditional waste disposal methods are
landfilling and incineration.   The modern and improved method for the same

0097–6156/93/0515–0072$06.00/0

process is now dedicated RDF. In some cases, the energy generated is used to produce steam for electricity generation which can be sold, and therefore constitutes a positive value. The problem here is that the mass-burn plant generates potentially polluting gaseous and solid residue effluents. In the gaseous effluent, dioxin has been one of the most elusive and worrisome pollutants and has caused the shutting down of a number of incinerator plants. There are other gaseous pollutants, including volatile refactory organics, chlorine containing compounds, and particulates from plastic and organic waste. The chemical and biological activity in the remaining solid ash residue from incinerators is also a problem which still requires landfilling or other methods of disposal. There is concern that leachates from incinerated ash will eventually contaminate the aquifers. Municipalities are also passing legislation forbidding the use of materials which do not degrade and tend to remain in long-term storage in the landfill, such as plastics. A number of communities are outlawing disposable plastic products and appear to be returning to paper bags and containers. Much effort is also going into developing biodegradable plastics. Whether this is a sound environmental solution is yet to be determined.

HYDROCARB Waste Process

The HYDROCARB Process offers a viable alternative to mass burn. The process was originally conceived for the purpose of processing our vast resources of coal to produce a clean carbon fuel (*1,2*). However, the process can operate as well with virtually any carbonaceous raw material and certainly a large fraction of MSW qualifies as a carbonaceous material. The process is new and unique and the products formed can be used primarily as premium clean fuels as well as for the commodity market. The process depends on two basic steps, (1) the hydrogenation of coal to form a methane-rich gas while leaving the ash behind and (2) the thermal decomposition of the methane-rich gas to form carbon black and hydrogen which is recycled. The excess hydrogen and oxygen from the co-products can be a hydrogen-rich co-product which can either be hydrogen, methane, methanol or water. Figure 1 shows a schematic flow with alternative feedstocks, coal, wood or MSW and with co-feedstock additions.

Figure 1 gives a schematic of the process listing various feedstock materials, additives and co-products. The process can be made very efficient because the only raw material used is the carbonaceous material and the energy required to operate the process is relatively small compared to the gasification process. The overall reaction is thermally neutral. The primary product is always carbon black which can be used as a clean burning fuel and can also supply the market for vulcanization of rubber for automotive tires, pigment for inks and paints and for lubricants. The co-product hydrogen-rich gas can primarily be used as a burner fuel and the methanol as an automotive fuel, or as a commodity chemical, or can even be converted to gasoline. The process is fundamentally different than mass-burn in that it operates in a reducing atmosphere rather than in an oxidizing atmosphere and it is run in a closed system under pressure. Temperature conditions are about the same or perhaps even somewhat lower than in mass-burn incinerator plants. Because of the elevated operating pressure and reducing atmosphere, no dioxin can be formed thermodynamically. All the oxygen contain-

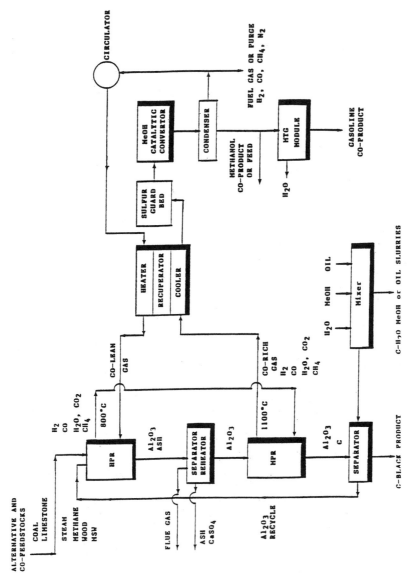

Figure 1. HYDROCARB process—Clean carbon from carbonaceous feedstock with methanol and gasoline co-product. Reproduced with permission from ref. 2.

ing organic material is reduced to carbon and methane and any metals that may be present in the waste are kept in their reduced state as opposed to mass-burn where the metals can become oxidized. The following describes how the process can be effectively used in processing MSW and the economic dynamics of the process.

The process can be used with either separated or non-separated waste. To simplify the example and avoid discussion of front-end costs, we will give examples of the process operating on separated waste. Thus, the main MSW feedstock is paper and plastic and we can include rubber tires for this example. Since paper is essentially produced from wood, the process can be represented by the following chemical stoichiometric formula, limiting the products to carbon and methanol.

$$CH_{1.44}O_{0.66} + 0.44CH_{1.5} + 0.14CH_4 = 0.92C + 0.66CH_3OH$$

paper + plastic + nat. gas black
methane

Notice that the formula for plastic contains only C and H, like rubber and methane. The oxygen containing material in paper is in the form of hemi-cellulose. The above equation is based on an assumed MSW composition such that the amount of plastic is 25% of the weight of paper. This can be changed for specific sites and the mass balance adjustment can be made by varying the amount of natural gas added. The gas can be purchased from the local gas company in the particular area where the waste is being processed. We now have to set the production capacity of the plant. Mass-burn incinerator plants have been built in the 2,000–3,000 T/D capacities in and around metropolitan areas. Of course, around New York, for example, it might be worthwhile building a 10,000 T/D or more of waste paper and plastic HYDROCARB plant. However, for this and generally more widespread applications, we will fix on a 3,000 T/D MSW processing capacity which would contain 2,400 T/D paper and 600 T/D plastic.

We now calculate that to run this plant, we have to add 226 T/D of natural gas from the natural gas pipeline company's distributing company. This natural gas is equivalent to 10.7 million SCF/D of methane, which must be purchased from the gas company. The separated MSW is thus co-processed with natural gas.

Economics

We now must estimate the capital investment of the plant. We can obtain this estimate by scaling down from a large plant we estimated in detail, operating on 25,000 T/D of coal. Because this is a volumetrically controlled process, we can scale it by the well known 0.6 power factor of capacity. The 25,000 T/D plant making carbon and methanol from coal is estimated to cost 800×10^6. Thus, the 3,000 T/D waste plant will cost:

$$800 \times 10^6 \times \left(\frac{3000}{25000}\right)^{0.6} = \$200 \times 10^6$$

We can now calculate a selling price for the carbon black fuel and methanol co-product. The financial parameters operating on the capital investment are as follows: capitalization 80% debt/20% equity, 20 yr depreciation, 11% interest on debt, 25% return on equity (ROE) and 38% tax on ROE before taxes. This results in a 21.9% annual fixed charge operating on the total capital investment.

We assume a high natural gas cost from the gas company of $5.00/MSCF which equals a cost of $0.119/lb CH_4. We then add operation and maintenance cost and the 21.9% fixed charges on the $200 million capital investment. We can now calculate the price (G) of the MSW value of the waste taken from the municipality, which can range from a negative value, in which case the community pays the processor to take the waste away, to a positive value in which case the processor pays the community to acquire the waste for processing. We shall first calculate a breakeven G price for the waste in $/Ton in Table I, assuming we obtain $5.00/MMBtu for the resulting fuel products.

Solving for G = $41.50/Ton; this is what the processor can afford to pay the town for taking the MSW for processing and while still obtaining a 25% return on equity.

The above is based on a fuel value for a C-methanol composition makeup mixture of 34.3% carbon in 65.7% methanol by weight. The plant produces 700,000 gal/Day of this C-methanol slurry which is equivalent to 11,000 Bbl/D of fuel oil equivalent.

If we assume the processor obtains the waste from the town free, so that G = $0/Ton, we can then calculate the selling price of $3.10/MMBtu for both *co-products* carbon and methanol. This is equivalent to $18.70/Bbl oil or $0.44/gal.

Now if the town pays the processor $25/Ton to cart the waste away (as some towns on Long Island have already done), then the selling price for carbon and methanol can come down to only $2.00/MMBtu which is equivalent to

Table I. HYDROCARB Waste Processing Plant[a]

Production Cost		$/Day
Waste Cost	=	3,000 T/D × $G/Ton
Nat. Gas = 0.119 × 226 × 2,000	=	53,000
Op. & Maint. = $\dfrac{3,000}{25,000} \times 120,000$	=	20,000
Fixed Charges = $\dfrac{0.219 \times \$200 \times 10^6}{328}$	=	$\dfrac{133,000}{206,000 + 3,000\,G}$

Thus,

$$206,000 + 3,000\,G = 0.9 \times (3,000 + 226\ \text{T/D}) \times \frac{22.9\ \text{MMBtu}}{\text{Ton}} \times \frac{\$5.00}{\text{MMTU}}$$

[a]Plant factor 90%, efficiency 90%, capacity 3,000 T/D, production capacity of fuel – 11,000 Bbl/D fuel oil equivalent.
SOURCE: Reproduced with permission from ref. 2

$12.00/Bbl fuel oil equivalent or $0.28/gal while maintaining a reasonable return on the investment equity.

At $2.50/MMBtu which is highly competitive with oil at $15.00/Bbl, the town would only have to pay $13.50/Ton to a processor to take it away.

The conclusion is that even at a waste capacity of 3,000 T/D and an investment of 200×10^6, the processor can sell the carbon and methanol as a clean burner fuel for domestic and industrial boilers, as well as for diesel and turbine engines at an economically attractive price. Additional return can be obtained by the processor selling the methanol and carbon at a higher price to the chemical commodity market so that the cost of waste disposal would even bring a profit to the town by selling the waste to the processor at a higher price.

The above preliminary assessment indicates that the HYDROCARB Process for the disposal of MSW is highly attractive and should be taken up for development on a fast track schedule. Because this process utilizes natural gas for co-processing waste in a reducing atmosphere, not only is the process environmentally acceptable but is potentially economically attractive and thus it should be worthwhile to develop this process in conjunction with a municipality that is generating the waste.

Literature Cited

1. Grohse, Edward W.; Steinberg, Meyer. *Economical Clean Carbon and Gaseous From Coal and Other Carbonaceous Raw Material;* Brookhaven National Laboratory, Upton, NY; BNL 40485, 1987.
2. Steinberg, Meyer. *Coal to Methanol to Gasoline by the HYDROCARB Process;* Brookhaven National Laboratory, Upton, NY; BNL 43555, 1989.

RECEIVED July 31, 1992

Chapter 7

Biomass-Fueled Gas Turbines

Joseph T. Hamrick

Aerospace Research Corporation, 5454 Aerospace Road,
Roanoke, VA 24014

Biomass fueled gas turbines offer a low cost means of
generating power in areas where fossil fuels are scarce
and too costly to import. This chapter presents results
of a ten year research and development effort to advance
biomass fueled gas turbines to commercial status. Aside
from trees the most promising sources of biomass are
sugar cane and sweet sorghum. Fuel alcohol can be produced from the sweet juices and grain for food is
produced by the sweet sorghum. As the gas turbine start-
up and shut-down effort is much less involved than the
start-up/shut-down of a steam turbine power plant, the
biomass fueled gas turbine is the preferred system for
meeting the peak electrical power demands of any system.
As CO_2 is extracted from the atmosphere during biomass
growth, and ultimate replacement through the gas turbine
combustion process is in exact balance, the entire process results in no net increase in CO_2 in the atmosphere.

Use of biomass to produce power in the five to thirty megawatt range
is attractive because of higher hauling costs associated with power
plants with higher outputs. The higher hauling costs result primarily from the low density of biomass. The ability to fuel gas turbine
power generating systems is economically advantageous due to the
quick start up and shut down capability. This capability allows re-
striction of operation to periods of peak usage for which higher
prices are paid for power. Steam power generation systems by comparison cost substantially more per installed kw output, and they cannot
be economically started up and shut down on a daily basis. Research
and development of biomass fueled gas turbine systems over a ten year
period has culminated in successful operation of an aircraft derivative gas turbine manufactured by the Allison Division of the General
Motors corporation. With the cooperation of the Tennessee Valley
Authority the power generating system was successfully integrated into their distribution grid for full demonstration of the system as a
small power producer using sawdust as the fuel. The Allison gas tur-

0097–6156/93/0515–0078$06.00/0
© 1993 American Chemical Society

bine served well in the research and development phase, but it is
economically marginal as a power producer in biomass fueled applications. As a result, a larger gas turbine produced by the General
Electric Company is the current selection for commercial use. A system using that engine is being prepared for installation at Huddleston,
Virginia, where a contract for sale of the output power to the Virginia Power Company has been negotiated. The R&D system which is
located at Red Boiling Springs, Tennessee will be upgraded for operation with the General Electric engine. Sugar cane bagasse, an alternative fuel to sawdust which has been tested in the R&D facility,
promises to be a major source of fuel in the future. Sweet sorghum
bagasse can serve equally well.

Sweet sorghum and sugar cane juices are readily converted to alcohol by yeast fermentation. Sweet sorghum can be grown throughout
the United States as well as the tropic and temperate zones of the
earth. These plants have the highest conversions of solar energy into
biomass of any of the species in the plant kingdom, substantially
greater than trees. With the use of bagasse as a fuel for gas turbines
in the generation of power, it is possible for the income from power
sales to reduce the cost of ethyl alcohol well below that for gasoline. The sorghum grain can be used for fermentation or food. The
high volume, high temperature exhaust gases from the turbine can be
used to concentrate the juice, make alcohol, dry the bagasse or generate steam for injection into the turbine. There is adequate heat
to concentrate the juice and dry the bagasse for year-round use
during the harvest period.

Growth of sugar cane and sorghum on the 66.4 million acres of
land taken out of production in the U.S. between 1981 and 1988 can
supply enough energy to generate 34 percent of the nonnuclear power
that was generated in 1986, enough to supply increased power demands
into the next century. At the current rates paid by Virginia Electric
Power Company for power generated with renewable fuels, 25.4 billion
gallons of alcohol can be produced from the profits earned on power
sales, enough to supply gasohol to the entire nation.

The system, which can be located at any point where there is a
power distribution line and a sorghum or sugar cane source, can provide jobs in the area and an alternative crop for farmers while sav-
ing billions of dollars on set-aside payments. At $20/barrel, approx-
imately $8 billion could be saved on the trade imbalance by the
reduction of oil imports by use of the set aside acres.

Background Information

Research on wood burning gas turbines was started by Aerospace Research Corporation in 1978. It culminated in the operation of an
Allison T-56 gas turbine power generating system at a facility located in Red Boiling Springs, Tennessee. Over two million dollars in
U.S. Department of Energy funds and a matching amount in private
funds were spent in carrying out the program. In addition, gas turbine
engines were furnished by the Air Force and Naval Air Systems Com-
mand. Results of the research and development effort are provided in
Reference 1. A schematic diagram of the system is provided in Figure
1 and a view of the R&D facility in Figure 2. The gas turbine, electric generator, and electrical switch gear configuration are the
same as that for the thousands of aircraft derivative gas turbine

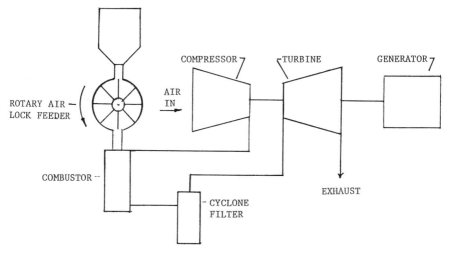

Figure 1.   Schematic Diagram of Biomass Fueled Gas Turbine
Power Generating System

Figure 2.   View of Facility at Red Boiling Springs, Tennessee

generator sets now in use throughout the world. The difference is in
the fuel system. The normal arrangement for the combustion chambers
of commercial gas turbines derived from those used on aircraft is the
same as that used on the aircraft, namely a configuration sandwiched
in between the compressor and turbine. Because of the slower burning
rate for biomass solids, the greater difficulty in feeding the fuel
into the combustion chamber, and the need for cleaning the ash from
the combustion products, a different arrangement is necessary. To
meet that requirement, the space between the compressor and turbine
must be adequate to provide room for ducting to take air off the compressor for movement to an offset combustion system and room for
return ducting to the turbine. Both the Allison T-56 gas turbine used
in the R&D system and the General Electric LM1500 gas turbine meet
that requirement. Meeting that requirement allowed rotating components in both gas turbines to remain unchanged, thus providing the
same mechanical reliability for the biomass fuels as for liquid fuels
without prohibitively expensive modifications.

At the outset of the R&D program reports on work involved with
solids fueled gas turbine systems (2-5) were thoroughly reviewed. It
was determined from research work with coal performed by the British
at Leatherhead, England (2) that single large cyclones operating
alone or in series would be best for ash cleaning. The combustion
chamber design was based upon information provided by the Australians
in Reference 3. An indication of problems that might be encountered
with biomass were provided by the report from Combustion Power Company, Inc. of California (4). From experience gained by Yellot et al
(5) in coal fired gas turbine research in the United States on solids
feeding it was learned that attempts to use lock hoppers to feed
solids into a combustor resulted in erratic operation of the gas
turbine. As a result, rotary air lock feeders, which were used in
all of the systems reported in References 3, 4, and 5, were selected
as the feeders in the R&D program.

Operational difficulties which resulted in learning curves
peculiar to the system such as wood processing, conveying, drying,
combustion ash removal, engine starting, synchronization with the
TVA power distribution grid, and development of emergency procedures
are covered in Reference 1. Feeding a pulverized solid into a high
pressure chamber and dealing with turbine blade fouling presented the
greatest challenge. An anticipated problem that was most feared at
the outset, eroding of the turbine blades, never materialized. In
over 1500 hours of operation, no erosion has been detected. The measures taken to resolve the two problems and the approach taken with
the General Electric LM1500 gas turbine in meeting the problems are
presented.

Modern aircraft engines which require very high power to weight
ratios are designed for high turbine inlet temperatures and high compressor discharge pressures. As turbine blade cooling techniques,
advanced materials and more sophisticated design methods have become
available the pressure ratios and allowable turbine inlet temperatures have increased to high levels. As a result, the modern aircraft derivative gas turbines are less suitable for operation with
biomass than the earlier models. The current need for low turbine
inlet temperature and low combustor pressure with biomass makes
earlier models more compatible. The LM1500 gas turbine fits well
into the biomass picture.

The Rotary Air Lock Feeder

The rotary air lock feeder is also referred to as a rotary valve. A schematic view of a rotary feeder is shown in Figure 1. Referring to Figure 1, the tips and sides of the vanes are fitted with seals that compartmentalize particles fed into a low pressure sector for movement around to a zone of high pressure and thence into the combustor. A major effort was directed toward development of long lasting sealing methods and materials. Sawdust is an extremely abrasive material that requires special techniques that were developed in the program. To meet the 130 psig requirement of the R&D installation two air lock feeders operating in series proved adequate. To provide conservative design margins, it is planned to use two feeders in series for the 90 psig pressure requirements of the Huddleston installation as well as in succeeding installations up to 6000 kw.

Turbine Blade Fouling

The primary problem with coal fired turbines was erosion of the turbine blades. A secondary problem was fouling of the blades. In work performed by the Coal Utilisation Research Laboratory at Leatherhead, England (2) it was determined that single cyclones in series adequately cleaned the ash from the combustion gases to prevent erosion. Therefore, it was decided to use only single cyclones in the wood burning program. As a result, there has been no erosion of the turbine blades in the more than 1500 hours of operation with the gas turbines used in the R&D program. In the R&D program performed by the Australians (3) on brown coal it was found necessary to limit the turbine inlet temperature to $1200^{O}F$ to avoid deposition of ash on the turbine blades. In the R&D performed at Leatherhead, England with stationery blades there was no significant deposition at $1450^{O}F$ after 1000 hours of operation with black coal. In tests with pine sawdust in early operation at Roanoke with a small Garrett turbine no significant deposition occurred at $1450^{O}F$ in 200 hours of operation. In tests with the Allison T-56 at Red Boiling Springs it was found necessary to periodically clean the turbine blades with milled walnut hulls when firing with a mixture of oak and poplar sawdust at $1450^{O}F$ turbine inlet temperature. Above $1450^{O}F$ the particles adhered to the blades and could be removed only by scraping. The $1248^{O}F$ turbine inlet temperature needed to produce 4000 kw with the LM1500 gas turbine in the Huddleston installation is well below any problem zone for disposition with sawdust.

Discussion of LM1500 Gas Turbine Performance And Factors Favoring Its Selection

When the advancement was made from the Garrett 375 kw gas turbine to a larger engine, the Allison T-56-9 gas turbine selection was made on the basis of its perceived easy adaptability to the system and the availability of used engines from the U.S. Air Force. As the R&D program advanced, it became clear that the turbine inlet temperature would have to be restricted to $1450^{O}F$ to avoid excessive turbine blade fouling. The turbine inlet temperature of the T-56-9 is 1700° F at its normal rated overall electrical output of 2332 kw. With a

1450°F turbine inlet temperature the output drops to 1500 kw, a
value too low for economical operation.

   A search for a more suitable gas turbine from standpoints of
availability, adaptability to wood fueling, and electrical output led
to selection of the General Electric J-79 gas generator and companion
power turbine. The combined gas generator and power turbine was
given the designation LM1500 by General Electric. For aircraft pro-
pulsion the hot gases leave the engine at high velocity, propelling
the aircraft forward. For use in power production the hot gases are
ducted to a power turbine. A favorable feature of a two shaft arrangement, such as this one, is that the gas generator can operate
efficiently at part load by adjusting its speed downward while the
power turbine operates at the required constant speed for power generation. The compressor efficiency is high over a broad range. This
is made possible by adjustment of variable stators in the first six
stages of the compressor. By adjustment of the stators to match the
compressor speed and air flow, rotating stall is avoided and good
compressor efficiency is maintained. Rotating stall is a phenomenon
associated with flow separation on the compressor blades as the angle
of attack on the blades increases with changes in rotative speed and
air flow. Compressor efficiency over the speed range results in
economical operation over a wide range of power production. Detailed
information on turbine inlet temperature, compressor discharge pressure, and wood feed rate as a function of power output was derived
from General Electric specification MID-S-1500-2.

Turbine Inlet Temperature. Figure 3 shows a straight line relationship between turbine inlet temperature and generator output. This
characteristic provides a significant amount of latitude in operation
with untried species of plants or sources of fuels such as clean
waste. For example, it can be safely predicted that in the worst
case the turbine inlet temperature of 1200°F required for a 3400 kw
output will not result in excessive or difficult to clean accumula-
tions on the turbine blade. Minimum performance guarantees would be
warranted in such cases. With most wood species a 7000 kw output
probably can be tolerated.

Compressor Discharge Pressure. Figure 4 shows a straight line relationship between compressor discharge pressure and generator output.
The primary concern with pressure is the feeding of solid fuel into
the combustion chamber. The demonstrated maximum sustained pressure
in the R&D system is 130 pounds per square inch. Thus, the ability
to sustain feeding of sawdust from 3000 kw to approximately 7500 kw
is assured. The pressure required to produce the 4000 kw projected
for the production facility now being prepared for installation at
Huddleston, Virginia is only 90 psig.

Wood Feed Rate. The wood feed rate in Figure 5 is based upon a heat
value of 8,200 Btu/lb. for sawdust. The heat value ranges from 8100
Btu per pound for oak to 8,600 Btu per pound for yellow pine. Green
sawdust as delivered from the mill averages approximately 45 percent
water content. Trailers 40 ft. long normally deliver on the order of
25 tons of green sawdust per load. For the 4000 kw output projected
for the Huddleston facility five trailer loads per day will be required. Shelter for approximately fifty truck loads will be needed
to assure continued operation in the winter months.

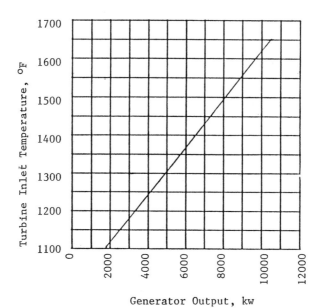

Figure 3. Plot of Turbine Inlet Temperature Versus Generator
Output With G.E. LM1500 GAS TURBINE at 1000 ft. Altitude and
Compressor Inlet Air at 70°F

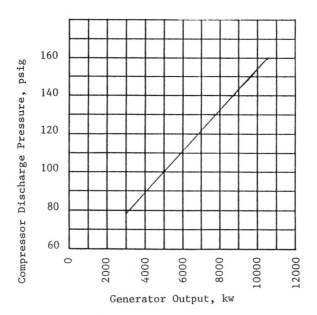

Figure 4. Plot of Compressor Discharge Pressure Versus
Generator Output With the G.E. LM1500 GAS TURBINE at 1000 ft.
Altitude and Compressor Inlet Air at 70°F

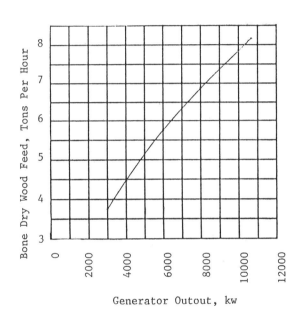

Figure 5. Plot of Wood Feed Versus Generator Output With
G.E. LM1500 GAS TURBINE at 1000 ft. Altitude and Compressor
Inlet Air at 70°F

Engine Durability

A question frequently arises as to the life of an aircraft derivative gas turbine in stationary power applications. The answer is that the lower power output and lower turbine inlet temperature that are projected for this application make for very favorable longevity for the LM1500 gas turbine. A twelve year life or greater before overhaul is predictable. The gas generator in the stationary application is never exposed to the extreme power requirements and high turbine inlet temperatures that exist during airplane take off. The primary requirement for long engine life in stationary applications is adequate filtration of the air entering the compressor. While longer term operation may prove otherwise, no erosion or corrosion of the turbine has been detected due to ash or dirt escaping the cyclone filter or from chemicals in the combustion products.

Sugar Cane and Sweet Sorghum Fuels

Sweet sorghum which is highly drought resistant can supply two to three times as much fiber energy per acre as trees in some areas in addition to the sugar produced for alcohol and grain for food. The yield from sugar cane, in the areas where it can be grown, is even higher than for sweet sorghum. A further advantage is that there is no stigma attached to its use as a fuel, as there is with trees. This renewable fuel will result in a zero net increase in carbon dioxide. Based on the published research results (6) for sweet sorghum, the 66.4 million acres taken out of production between 1981 and 1988 can supply the energy to generate 34 percent of the nonnuclear power generated in 1986 in the U.S. The annual payment for setting land aside is estimated to be over $5 billion. Much more additional acreage can be easily devoted to sorghum as an alternative crop. Besides providing fuel for electric power the grain and sugar can produce in excess of 25.4 billion gallons of ethanol which equals fifteen percent of the energy supplied from imported oil. Intensive cultivation of sugar cane and sorghum in states bordering the Gulf of Mexico can result in tripling these outputs. Research results from Reference 6 are shown in Table 1.

Table 1. Maximum Sweet Sorghum Yields at 8 Sites in 1978 & 1979

Location	Maximum Yields, 1978–1979				Wray Cultivar Yields 1979	
	Biomass		Total Sugars		Biomass	Total Sugars
	Metric Tons Per Hectare					
	1978	1979	1978	1979		
Baton Rouge, LA	28.8	31.7	8.5	11.19	16.2	8.8
Belle Glade, FL	40.5(a)	12.0	13.2(a)	5.5	34.8(b)	13.2(b)
Columbus, OH	22.2	18.5	6.5	3.1	17.5	3.1
Fargo, ND(c)	--	12.5	--	2.9	12.5	2.7
Lincoln, NE(c)	--	19.3	--	6.3	16.0	6.3
Manhattan, KS(c)	--	24.7	--	4.0	19.2	3.9
Meridian, MS	22.4	28.8	7.7	8.6	28.8	8.6
Weslaco, TX	30.5	30.0	9.0	6.5	18.0	5.7

(a) This value is the sum of two crops of sweet sorghum produced during the 1978 growing season.
(b) This value is from 1978 results since Wray was not grown in Florida during 1979. Wray was double-cropped in Florida in 1978.
(c) This site was not included in sweet sorghum experiments in 1978.

It is indicated by the results shown for Belle Glade, Florida
that in many regions of the earth two crops per year can be grown.
The versatility of the biomass fueled gas turbine system in connection with disposal of waste and in cogeneration is currently undergoing evaluation.

Cogeneration Applications

The temperature of the exhaust gases from the biomass fueled LM1500
gas turbine ranges from 620°F to 685°F depending upon the turbine inlet temperature that can be tolerated by the biomass material in use.
In the case of the General Electric LM1500 gas turbine the exhaust
gas flow ranges from 125 to 135 lbs. per second. At 620°F and 125
lbs. per second, and cooling to 125°F the exhaust gases would yield
approximately 55 million Btu per hour, the equivalent of approximately
460 gallons of gasoline per hour. The heat thus generated can be
used in drying lumber, heating buildings, chemical processing as with
ethanol production, or injection of steam into the turbine for increased efficiency and power. Ethanol can be produced directly from
fermentation of the sweet juice pressed from sugar cane or sorghum or
from conversion of cellulosic feed stock to alcohol through hydrolysis and fermentation as discussed in References 7 and 8. For
example, bagasse from either sugar cane or sweet sorghum can be
burned directly in the gas turbine or processed for production of
ethanol. In the latter case approximately 26 percent of the bagasse
winds up as high heat content residue that can be burned in the gas
turbine.

Alcohol Production From Sweet Juice. Approximately 15 percent of the
exhaust gases will be needed to dry the bagasse for feeding into the
combustion process. The remainder is available for drying the
bagasse for storage and alcohol production. In some geographical
areas it may be necessary to use bagasse to supplement the residue
from the hydrolysis process to provide the needed fuel for the gas
turbine.

Systems Producing Ethanol From Biomass. Biomass from municipal
waste, logging residue or biomass plantations can be converted to
ethanol by way of the processes addressed in References 7 and 8.
According to Reference 8 lignocellulosic materials can be fraction-
ated into cellulose, hemicellulose, lignin and extractives by a
number of different technologies including solvent extraction and
various degrees of steam-aqueous treatment. The authors of Reference
8 point out the environmental advantage of steam-aqueous treatments
for which the water solvent is widely available at low cost with no
environmental impact. Fractionation is followed by enzymatic treatment of the carbohydrates toward production of ethyl alcohol and
burning of the noncellulosic components to produce power and heat.
There has been no economic assessment of a system involving direct
conversion to ethanol with subsequent burning of the residue in gas
turbines.

Steam Injection for Increased Power and Efficiency. Steam injection
for increased power was reported by Australian investigators (3) as
early as 1958. Steam injection was shown to increase power by 64
percent and efficiency by 25 percent for the case that was cited.
Similar results for power output were achieved with water injection
in tests with the Allison T-56 gas turbine at Red Boiling Springs,
Tennessee. It was determined that water from the available sources
at Red Boiling Springs would have to be deionized for a long term
operation. A major consideration in use of either water or steam
injection is the availability of water, as the water or steam that is
injected is lost to the atmosphere. For example, to increase the
output of the LM1500 gas turbine from 6 to 9 mw will require on the
order of 90 gallons of water per minute. Whether to use water or
steam will depend upon the cost of fuel. Steam does not require an
increase in the amount of fuel as heat from the exhaust gases produces the steam. Direct water injection requires additional fuel.

Summary

The General Electric LM1500 gas turbine has been chosen for use in
the wood burning power production system because of its highly compatible performance characteristics, the ease with which it can be
mechanically adapted to the system, and its ready availability.
Salient points are as follows:

1. The 4000 kw power output projected for the production system
being readied for installation at Huddleston, Virginia can be a-
chieved with a 1250°F turbine inlet temperature and compressor discharge pressure at 90 psig. Both are well below the 1450°F turbine
inlet temperature and 130 psig compressor discharge pressure found
acceptable in the R&D program.

2. Power outputs up to 7500 kw can be achieved with oak sawdust
while remaining below the 1450°F turbine inlet temperature and 130
psig compressor discharge pressure found acceptable in the R&D program.

3. There is adequate distance between the compressor and turbine to adapt the engine to the external burner required for wood
and other biomass fuels.

4. J-79 gas generators are readily available on the overhaul
and used market. New power turbines are available from manufac-
turers. In addition, a limited number of serviceable complete
LM1500 sets are available for immediate use.

5. Because of the high heat content of the exhaust gases, the
biomass fueled gas turbine can serve in a wide range of cogeneration
systems involving use of cellulosic materials.

6. Both Red Boiling Springs and Huddleston facilities are
ideally located for demonstration of combined electrical power and
fuel alcohol production from sweet sorghum.

References

1. Hamrick, J.T., Development Of Biomass As An Alternative Fuel
    For Gas Turbines, Report PNL-7673, UC-245 April 1991 DOE,
    Pacific Northwest Laboratory, Richland, Washington.
2. Roberts, A.G., Barker, S.N., Phillips, R.N. et al, Fluidised
    Bed Combustion, NCB Coal Utilisation Research Laboratory Final
    Report, U.S. Department of Energy Report FE-3221-15(a), 1980.
    National Technical Information Service, Springfield, Virginia.
3. F.L. McCay, Chairman, The Coal-Burning Gas Turbine Project,
    ISBN 0642001499, 1973. Australian Government Publishing Service,
    Canberra, NSW, Australia.
4. Summary of The CPU-400 Development, Staff, Combustion Power
    Company, Inc. U.S. Environmental Protection Agency Report on
    Contracts 68-03-0054 and 68-03-0143, October 1977. National
    Technical Information Service, Springfield, Virginia.
5. Yellot, J.I. et al, Development of Pressurizing, Combustion,
    and Ash Separation Equipment For a Direct-Fired Coal-Burning
    Gas Turbine Locomotive, ASME Paper 54-A-201, 1954. American
    Society of Mechanical Engineers, Washington, D.C.
6. Jackson, D.R., Arthur, M.F. et al, Development of Sweet
    Sorghum As An Energy Crop, Report BMI-2054 (Vol. 1) May, 1980.
    Battelle Columbus, Columbus, Ohio.
7. Lynd, Lee R., Cushman, Janet H., Nichols, Roberta J., and
    Wyman, Charles E., Fuel Ethanol From Cellulosic Biomass,
    Science, Vol. 251, 15 March 1991.
8. Heitz, M., Capek-Menard, E., Koeberle, P.G., Gagne, J.,
    Chornet, E., Overend, R.P., Taylor, J.D., and Yu, E.,
    Fractionation of Populus Tremuloids at the Pilot Scale:
    Optimization of Steam Pretreatment Conditions using the
    Stake II Technology, Biosource Technology 35 (1991) 23-32,
    Elsevier Science Publishers Ltd., England.

RECEIVED July 27, 1992

Chapter 8

Transition Metals as Catalysts for Pyrolysis and Gasification of Biomass

L. A. Edye, G. N. Richards, and G. Zheng

Wood Chemistry Laboratory, University of Montana, Missoula, MT 59812

The addition of Fe or of Cu ions to lignocellulosics such as wood or newsprint specifically catalyzes the pyrolytic formation of levoglucosan (LG) (up to 32% based on cellulose). The yield of charcoal is also increased (*e.g.* to 50% at 330°C). The ions may be added by ion exchange *via* the hemicelluloses, or by sorption of acetate salts from aqueous solution. The sulfate salts catalyze formation of levoglucosenone (LGO) in addition to LG. These effects probably involve complexing of the transition metal cations with lignin. The copper-doped charcoals obtained at ca. 350°C have been shown to contain elemental Cu, which can act as catalyst for clean gasification, either in nitrogen (to yield predominantly CO, CO_2 and hydrocarbons) or in an oxidizing gas.
When air is used to gasify the catalyzed charcoals a so-called "jump phenomenon" is often observed in the dependence of apparent gasification rate on gasification temperature. This effect may involve an increase of two orders of magnitude in rate for an increase of 5-10° in temperature when the rates are measured by TG. We have now shown that the "jump phenomenon" is partly, but not entirely due to an increase in sample temperature above reactor temperature during gasification. This increase is probably associated with exothermic oxygen chemisorption on the char and occurs on admission of air and on exposure of new reactive sites during gasification. Procedures have been designed to minimize errors from such effects on measurement of gasification rates by TG.

In the late 1980's the combustion of wood, agricultural crop residues and municipal solid waste to supply residential heat, process heat and steam, and electricity comprised about 4% of the total annual U.S. energy requirement (*1*). Although this contribution by simple combustion of biomass is significant, the potential of more technologicaly advanced thermochemical processes (*viz.* pyrolysis, liquefaction and gasification) promises an even greater contribution to future energy requirements. Consequently, thermochemical conversion of biomass to energy and useful chemicals continues to be a topic of active research.

The predominant approaches to thermochemical conversion of biomass involve either pyrolysis to oils (with or without hydrogenation) or alternatively direct gasification to combustible gases. These two approaches involve "overlapping" chemistry. The production of pyrolysis oils also generates chars, which are potential substrates for subsequent gasification. Direct gasification of biomass produces gases that may include both primary pyrolysis products and secondary reaction products. For example water, a pyrolysis product, may further react with the char to produce water gas. The major disadvantage of direct gasification is the inevitable co-production of tars, phenolics and acids. Although washing the gases removes these co-products (and produces gases for clean combustion), the composition of the washed gases is complex and they have only limited use for synthesis purposes.

We have adopted the approach that in some circumstances the optimum route to utilization of biomass may be to pyrolyze at moderate temperatures and then separately to gasify the resulting char, *e.g.*, with CO_2 to CO or with H_2O to CO + H_2. Hence, we envision a two stage process for the conversion of lignocellulosic waste (such as sawdust or newsprint) into useful chemicals and simple gases. In the first stage, pyrolysis at relative low temperatures produces a charcoal, a condensible organic liquid (pyrolysis tar) and a noncondensible low Btu fuel gas. Obviously, the process would benefit by the formation of useful chemicals in the pyrolysis tar in relatively high yields. In the second stage, the charcoal which contains a major portion of the energy content of the feedstock is either pyrolyzed at high temperatures, or gasified to produce simple gases. Since the gasification is free from tar forming reactions the product gases are cleaner and have greater value for synthesis (*e.g.* of methanol). The energy balance and product gas composition would be dependent on the composition of the reactant gas (*2,3*).

It is well known that pyrolysis of pure cellulose under vacuum produces a pyrolysis tar containing 1,6-anhydroglucose (levoglucosan; LG), in a relatively high yield (up to 66%) (*4*). The production of LG from vacuum pyrolysis of wood has also been reported (*e.g. 5,6*), but the pyrolysis of wood normally gives very low yields of LG relative to the cellulose content of the wood. However, we have recently reported higher yields of LG from wood pyrolysis after prior removal of the indigenous metal ions bound to the hemicellulose in the wood by washing with very dilute acid (*6,7*). In fact, numerous publications from this laboratory have described the catalytic influence of both indigenous and added metals and metal ions on pyrolysis and gasification of biomass (*6-19*). These effects have been studied by the addition of salts to wood (*e.g. 10*) and also by exchanging the indigenous metal ions in wood for other ions (*e.g. 11*). In the latter case the metal ions are held as counterions to the uronic acid constituents of the cell wall and can be controlled by ion-exchange procedures (7).

We have recently (17) studied the effect of individual metal ions (ion exchanged into wood) on gaseous products from pyrolysis of wood using coupled TG/FTIR. These studies showed that K but not Ca, acts as a catalyst in pyrolysis reactions resulting in formation of CO_2, CO, and formic acid (especially from polysaccharides), acetic acid (from hemicellulose), and methanol (from lignin). This type of study has now been extended to include a wider range of metal ions, and to embrace newsprint as well as wood with the aim of optimizing anhydrosugar yields in pyrolysis tars. Since the major motivation in this work lies in the thermochemical utilization of lignocellulosic waste, we have especially included metal ions which are known to increase char yield and to act as catalysts in char gasification.

The thermochemical conversion of biomass chars to synthesis gas and the catalytic effect of both indigenous and added metals in gasification of such chars have received relatively little attention compared to the similar studies on coal. Also, several studies have been made of the catalytic effects of copper and its salts in accelerating the oxidation of graphite (20-23). By comparison, to the best of our knowledge, there have been no prior reports on copper catalysis of biomass char gasification.

In two separate studies (15,16) of the influence in gasification of indigenous cations of biomass, we have been able to show that the rate of gasification in CO_2 of a range of biomass materials correlates well with the total metal ion content of the biomass, provided that biomass materials with high silica are omitted from the correlation. In such studies the chemical and physical heterogeneity of biomass presents problems in obtaining a uniform representative sample. Furthermore, most biomass materials are poor thermal conductors and hence during the course of thermal conversion of a biomass sample (a complex interaction of heat and mass transfer processes via a porous surface) even small particles may exhibit temperature gradients. For these reasons it is appropriate to model biomass gasifications on more uniform and definable carbon sources (e.g. CF11 cellulose and carboxymethyl cellulose). Even municipal solid waste, a highly variable material, contains on average ca. 50% cellulosic materials (43% paper, 10% yard waste, 10% glass, 7% metals, 15% food and 5% plastic) (24). The use of these model carbon sources also allows particluarly exact control of the distribution and concentration of metal species in the substrate. Thus we have utilized carboxymethylcellulose (CMC) for incorporation of ion exchanged cations and cellulose fibers (CF-11) for incorporation of salts by sorption from aqueous solution.

Here we are specifically concerned with air gasification of the chars doped with metal ions that result from pyrolyses designed to optimize production of LG and LGO (i.e. Fe and Cu-doped chars). In the case of coal (Spanish lignite char), copper is reported to be the most effective catalyst of the carbon-air reaction of the first transition series of elements (25,26). The same authors report (for air gasification of Cu-doped lignite char) a dramatic jump from a region of low reactivity and high apparent activation energy to a region of high reactivity and low apparent activation energy for a 5°C temperature increase (*27*). In fact, we have also observed this jump phenomenon for copper catalysis (*19*) and for iron catalysis (*28*) in the air gasification of cellulosic chars. For this study we utilized the model carbon sources described above. These Cu-doped cellulosic fibers were then pyrolyzed to chars for gasification studies. We were able to report "jump temperatures" (Tj, defined by Moreno-Castilla *et al.* (*27*) as the lowest temperature of the high reactivity region) for ion-exchanged Cu CMC and sorbed Cu salt CF-11 samples. Part of this chapter reports our continued study of the jump phenomenon.

Experimental

The ion-exchanged wood (milled cottonwood sapwood, *Populus trichocarpa*) and carboxymethylcellulose and salt-sorbed samples were prepared as described earlier (*17,18*). Newsprint samples were prepared from the Wall Street Journal, macerated with deionized water (or salt solutions for salt-sorbed newsprint) in a Waring blender, filtered and air-dried to a mat ca. 2 mm thick. The mats were cut into cubic pellets (*ca.* 2 mm^3) before pyrolysis. Metal ion contents were measured by inductively coupled argon plasma spectrometry (see Tables I to IV).

Pyrolyses were carried out at 2 Torr under flowing nitrogen as described previously (*29*) with tars condensed at room temperature and "distillate" condensed at -50°C. LG contents of tars were determined by GLC of tri-*O*-methylsilyl ethers (*29*) and the compounds in the second condensate (other than water) were determined by ^{1}H NMR (*30*). The yield of (LGO) in the second distillate was determined by relating the integrated signal for the C-1 hydrogen (5.31 ppm, s, 1H) (*31*) to the internal standard (2-methyl-2-propanol; 1.21 ppm, s, 9H).

The thermogravimetry system used to measure gasification rates has been described (*19*). The samples were first pyrolyzed at a heat treatment temperature (HTT) for 15 min in nitrogen *in situ* (80 mL min^{-1}), and subsequently gasified (gasification temperature = GT) in air (80 mL min^{-1}, 22% O$_2$). The TG balance was purged with helium (20 mL min^{-1}). The temperature program and weight loss curve of a typical gasification experiment are shown in Figure 1.

The apparatus for measuring differences in sample and furnace temperatures is described elsewhere (*32*). Essentially, a thermocouple was placed in the furnace near to the sample (similar to TG temperature measurement) and a second thermocouple embedded in the sample. The temperature program and gas flow control was similar to a typical gasification, and during the course of the experiment the two thermocouple readings were compared.

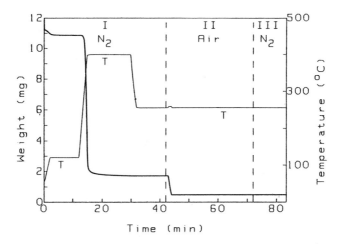

Figure 1.   CuCMC-1 Gasification in air; HTT 400°C, GT 255°C.
Reproduced with permission from reference 19.   Copyright 1990
Butterworth-Heinemann Limited.

RESULTS AND DISCUSSION

Pyrolysis of Ion-exchanged Wood. The yields of char, tar, and distillate from the pyrolysis of ion-exchanged wood are shown in Table I. The alkali metal and calcium wood samples showed increased char formation, low tar and high distillate yield compared to acid-washed wood. All other ion-exchanged wood samples showed char and tar yields similar to acid-washed wood, with the transition metals especially favoring tar formation. The analysis of the tar for LG shown in Table I indicates that the tars from Li, K and Ca wood samples were not only obtained in reduced yield, but also contained a lower proportion of LG than tar from acid-washed wood. All other metal ions produced yields of LG above 10.5% compared with 5.4% from acid-washed wood. In other words, wood samples ion-exchanged with transition metals gave higher yields of a cleaner tar.

Table I. Vacuum pyrolysis of ion-exchanged wood

Sample	Metal ion[a]	Char[a]	Tar[a]	Distillate[a]	LG[a]
Acid-Washed Wood	----[b]	19	43	16	5.4
Li	----	19	25	44	0.7
Mg	0.08	16	45	29	10.5
K	0.20	22	21	40	0.4
Ca	0.14	20	32	27	4.1
Mn[II]	0.19	15	44	29	10.6
Fe[II]	0.37	15	50	30	15.8
Co[II]	0.20	15	49	29	12.4
Ni[II]	----	14	56	25	13.0
Cu[II]	0.19	17	54	26	13.5
Zn	0.26	16	45	31	11.4
CF-11 cellulose	----	13	67	14	41.5
CuAc$_2$/CF-11[c]	----	24	51	14	31.0

[a]% weight based on dry ash free wood. Pyrolysis at 350°C/30 min.
[b]Not determined.
[c]Cupric acetate sorbed on CF-11 (not ion-exchanged).

The best yield of LG observed was 15.8% from the iron(II)-exchanged wood sample. Assuming that LG is derived only from cellulose and that the latter represents 50% of the wood (*33*), then the conversion of cellulose to LG was 31.6%. This yield may be compared with 41.5% obtained from pure cellulose (CF-11 cellulose).

The mechanism whereby the yield of LG is increased by the presence of these metal ions is not known. However, since cupric acetate sorbed in pure cellulose does not catalyze the formation of LG, we tentatively conclude that catalysis of LG formation in wood may involve some interaction with lignin. The low yield of LG from original wood (0.4%) is probably associated with two inhibiting factors, *viz.* indigenous metal ions (especially K and Ca) and lignin. When the metal ions are removed by acid washing, the LG yield increases to 5.4% (*i.e.* 10.8% based on cellulose), but the lignin effect presumably holds the LG yield well below that from cellulose (41.5%). The presence of some metal ions, especially the transition metals, may decrease the interference by lignin in the conversion of cellulose to LG.

Pyrolysis of Wood Sorbed with Salts. Since the improved yields of LG from wood pyrolysis induced by the presence of added metal ions are of considerable interest (in connection with thermochemical utilization of biomass), this study was extended to determine whether it is necessary to remove the indigenous cations in the wood, and whether the "beneficial" metal ion can be added more simply and economically as salts by sorption rather than by ion-exchange. This study concentrated on cuprous and ferrous salts because these ions were most effective in increasing LG and charcoal yields when ion-exchanged, and also because the chars from such pyrolyses are likely to contain elemental Cu or active Fe species which are known to be catalysts of gasification reactions (*19*).

The products of pyrolysis of wood sorbed with cupric acetate, ferrous acetate and ferrous sulfate are shown in Table II.

Table II. Vacuum pyrolysis of wood sorbed with salt solution

Sample[a]	Metal ion[a]	Char[a]	Tar[a]	Distillate[a]	LG[a]
Original wood		19	25	24	0.4
Acid-Washed (AW)		15	43	16	5.4
AW/CuAc$_2$	0.45	29	38	---[b]	11.2
AW/CuAc$_2$	0.74	30	40	---	11.2
AW/CuAc$_2$	1.10	20	42	---	12.5
CuAc$_2$	0.43	15	43	21	5.9
AW/FeAc$_2$	3.34	22	24	26	7.5
FeAc$_2$	3.18	28	19	28	5.2
AW/FeSO$_4$[c]	1.43	38	18	25	6.1

[a]% weight based on dry ash free wood. Pyrolysis at 350°C/30 min.
[b]Not determined.
[c]Pyrolysis at 300°C/60 min.

The addition of cupric ion by sorption of the acetate salt in acid-washed wood (to 0.45% Cu) is effective in increasing LG yield to 11.2%; higher concentrations of copper did not significantly improve the LG yield. When the indigenous ions (predominantly K and Ca) are not removed by acid washing before addition of the cupric acetate, the improvement in LG yield is much less marked. The indigenous cations negate some of the catalytic influence of the

Cu. The results with sorbed ferrous acetate were similar, although the LG yield with the salt sorbed in the acid-washed wood (7.5%) was considerably less than for the corresponding ion-exchanged wood (15.8%).

The pyrolysis of wood sorbed with ferrous sulfate was studied as an example of an anion likely to remain in the pyrolyzing solid and to generate acid conditions (acetate ions are lost from the solid at the pyrolysis temperature). Acid-washed wood containing ferrous sulfate yielded 6.1% LG and an additional 3.0% levoglucosenone (1,6-anhydro-3,4-dideoxy-β-D-glycero-hex-3-enopyranos-2-ulose, LGO), which was found in the distillate. LGO is a known product of acid-catalyzed cellulose pyrolysis (*34*) and its formation was observed only in the presence of the sulfate anion. The yields of LGO from wood, shown in Table II (and not yet optimized) are greater than those commonly reported from pure cellulose or paper. This is particularly opportune in view of current interest in use of LGO as a versatile synthon.

Application to Newsprint. Newsprint comprises about one-third of the solid municipal waste in developed countries, it is likely soon to be excluded from landfill disposal and there is a limit to the proportion which can be recycled into paper. Thermochemical processes have the potential to account for the utilization of large amounts of waste newsprint. Our sample contained 24% lignin as determined by the method of Iiyama and Wallis (*35*). It was therefore expected, on the basis of the above experiments with wood, to be similarly amenable to the "beneficial" effects of ferrous sulfate in terms of increased LG and LGO yields.

Table III shows the influence of sorbed ferrous sulfate on products of pyrolysis in nitrogen at 400°C for 30 min. In the absence of added ferrous sulfate, pyrolysis of newsprint yielded 3.2% LG, but no LGO. Sorption of ferrous sulfate (to 2.08% $FeSO_4$) before pyrolysis increased the yield of LG to 16.6%, and also yielded 4.2% LGO. Char yield also increased from 15% for newsprint to 20% for $FeSO_4$ sorbed newsprint. Thus we have a procedure to generate chemical feedstocks (LG and LGO) from pyrolysis of newsprint, while simultaneously forming in increased yield a char which already contains an efficient gasification catalyst.

Table III. Vacuum pyrolysis products from newsprint (NP)[a]

Sample	Char	Tar	Distillate	LG	LGO
Original NP	15	44	32	3.3	ND
NP + 2.08% $FeSO_4$	20	43	37	16.6	4.2

[a]% weight based on dry ash free wood. Pyrolysis at 400°C/30 min.
ND Not detected.

Gasification of Cu and Fe Ion-exchanged CMC Chars. We have previously reported the effect of HTT and mode of addition of metal ions on the Tj of Cu-doped chars (*19*). Jump temperatures were determined in a series of gasification experiments where GT was successively lowered 5 °C until the low reactivity region was reached. The Tj for Fe-doped CMC char (HTT 400 °C, [Fe] 2.47% d.a.f.), determined by the same method, was 295 °C. Figures 2 and 3 show rate-time plots for the gasification in air of Cu- and Fe-doped cellulose chars (HTT 400 °C) at and below the jump temperature (Tj = 255 °C for Cu and 295 °C for Fe). Copper appears to be a superior catalyst over iron; the Tj of Cu is lower and the rate maximum (Rg(max)) is higher. In the case of Cu-doped chars, the initial rate of gasification increases *ca.* 160-fold for a 5 °C temperature increase at Tj.

In a series of experiments to determine the effect of Cu concentration on Tj we gasified Cu-doped CMC chars in a temperature program (HTT 400 °C, GT 200 °C + 5°/min). Table IV shows the effect of Cu concentration on the apparent Tj. Increasing Cu concentration in the range shown in Table IV effected a decrease in the apparent Tj but did not change Rg(max). We expected to measure Tj as the temperature at which the rate suddenly increased, and that the Tj of our CuCMC-1 sample would be as previously determined (*i.e.* 255 °C). However, the Tj of CuCMC-1 determined by this method was significantly different. We conclude that this jump in reactivity is also affected by the thermal history of the char.

Table IV. The effect of copper concentration on the apparent jump temperature[a]

Sample[b]	Cu in char (% d.a.f.)	Tj (°C)	Rg(max) (min⁻¹)
CuCMC-1	2.82	312	1.6
CuCMC-2	1.54	320	1.8
CuCMC-3	0.67	340	1.8

[a]For HTT 400 °C, GT 200 °C + 5°/min.
[b]Carboxymethylcellulose, ion exchanged with various loadings of Cu^{2+}.

Effect of Oxygen Chemisorption on Sample Temperature in the Furnace. It seemed likely, based on the observation of small spikes in the TG thermocouple reading on the introduction of air into the furnace (see Figure 1), that the TG apparatus may not be recording the true temperature of the sample during rapid gasification. In fact, when a *ca.* 20 mg CuCMC-1 char (HTT 400 °C) was gasified (GT 260 °C) with a thermocouple in contact with the char, the temperature of the char ran ahead of the furnace temperature soon after air was admitted (see Figure 4).

We also observed (*32*) that for calcium-doped chars prepared at relatively low temperatures (*e.g.* HTT 400 °C), the initial rate of gasification in

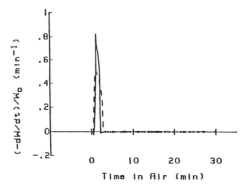

Figure 2. DTG curves for Fe and CuCMC at Tj. —— CuCMC-1 GT 255; − − FeCMC GT 300.

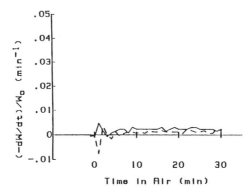

Figure 3. DTG curves for Fe and CuCMC below Tj. —— CuCMC-1 GT 250; − − FeCMC GT 250.

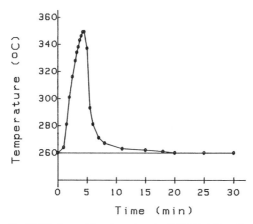

Figure 4. CuCMC-1 gasification in air (HTT, 400°C). ⊖ sample temperature; —— furnace temperature.

air (Rg(max) from TG) is extremely high for a short time. This effect was due to "run away" temperature increase associated with exothermic oxygen chemisorption. In this case the effect could be avoided by pre-sorption of oxygen below the gasification temperature. However, unlike calcium-doped chars, in the copper-catalyzed gasification the initial high Rg(max) peak at GT 260°C could not be eliminated by pre-sorption of oxygen at 200°C (see Figure 5). We conclude that exothermic oxygen chemisorption contributes to, but does not fully account for the jump phenomenon in copper catalysis of gasification.

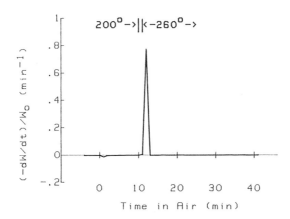

Figure 5.   CuCMC-1 Gasification with pre-sorption at 200°C.

Acknowledgments

The authors are grateful to H.-X. Qui, G.R. Ponder, M.P. Kannan and G. Devi for experimental assistance. This work was financially supported by the McIntire Stennis Project MONZ 8701 and by the Gas Research Institute Grant No. 5088-260-1639.

Literature Cited

1.   Walter, D.K.   In *Research in Thermochemical Biomass Conversion;* Bridgwater A.V.; Kuester, J.L., Eds.; Elsevier Appl Sci., London, 1988, pp 10-15.
2.   Walker, P.L.; Rusinko, F.; Austin, L.G. *Advan. Catal.* **1959,** *11,* 133-221.
3.   Brink, D.L.; Thomas, J.F.; Faltico, G.W.   In *Fuels and Energy from Renewable Resources;* Tillman, D.A.; Sarkanen, K.V.; Anderson, L.L., Eds.; Academic Press, New York, NY, 1977, pp 141-168.
4.   Essig, M.; Richards, G.N.; Schenck, E.   In *Cellulose and Wood Chemistry and Technology;* Schuerch, C., Ed.; John Wiley & Sons, New York, NY, 1989, pp 841-862.
5.   Shafizadeh, F.; Furneaux, R.H.; Cochran, T.G.; Scholl, J.P.; Sakai, Y. *J. Appl. Pol. Sci.* **1979,** *23,* 3525-3539.

6.   Ponder, G.R.; Qiu, H.-X.; Richards G.N. *App. Biochem. Biotechnol.* **1990,** *24/25,* 41-47.
7.   DeGroot, W.F. *Carbohydr. Res.* **1985,** *142,* 172-178.
8.   DeGroot, W.F.; Shafizadeh, F. *Carbon,* **1983,** *21,* 61-67.
9.   DeGroot, W.F.; Shafizadeh, F. *J. Anal. and Appl. Pyr.* **1984,** *6,* 217-232.
10. DeGroot, W.F.; Shafizadeh, F. *Fuel* **1984,** *63,* 210-216.
11. DeGroot, W.F.; Richards, G.N. *Fuel* **1988,** *67,* 352-360.
12. DeGroot, W.F.; Petch, G.S. In *Applications of Chemical Engineering Principles in the Forest Products and Related Industries;* Kayihan, F.; Krieger-Brockett, B., Eds.; American Institute of Chemical Engineers/Firest Products Davision, Tacoma, WA, 1986, pp 144-151.
13. DeGroot, W.F.; Richards, G.N. *Fuel* **1988,** *67,* 345-351.
14. DeGroot, W.F.; Osterheld, T.H.; Richards, G.N. In *Research in Thermochemical Biomass Conversion;* Bridgwater, A.V., Kuester, J.L., Eds.; Elsevier Appl. Sci.: London, pp 327-341.
15. DeGroot, W.F.; Kannan, M.P.; Richards, G.N.; Theander, O. *J. Agric. Food Chem.* **1990,** *38,* 321-323.
16. Kannan, M.P.; Richards, G.N. *Fuel* **1990,** *69,* 747-753.
17. Pan, W.-P.; Richards, G.N. *J. Anal. Appl. Pyrol.* **1989,** *16,* 117-126.
18. Kannan, M.P.; Richards, G.N. *Fuel* **1990,** *69,* 999-1006.
19. Ganga Devi, T.; Kannan, M.P.; Richards, G.N. *Fuel* **1990,** *69,* 1440-1447.
20. McKee, D.W. *Carbon* **1970,** *8,* 131.
21. McKee, D.W. *Carbon* **1970,** *8,* 623.
22. Patrick, J.W.; Walker, A. *Carbon* **1974,** *12,* 507.
23. Turkdogan, E.T.; Vinters, J.V. *Carbon* **1974,** *12,* 189.
24. Mallya, N.; Helt, J.E. In *Research in Thermochemical Biomass Conversion;* Bridgwater, A.V.; Kuester, J.L., Eds.; Elsevier Science Publishing Co., Inc.: New York, NY, 1988, 111-126.
25. Fernandez-Morales, I.; Lopez-Garzon, F.J.; Lopez-Peinado, A.; Moreno-Castilla, C.; Rivera-Utrilla, J. *Fuel* **1985,** *64,* 666.
26. Moreno-Castilla, C.; Rivera-Utrilla, J.; Lopez-Peinado, A.; Fernandez-Morales, I.; Lopez-Garzon, F.J. *Fuel* **1985,** *64,* 1220.
27. Moreno-Castilla, C.; Lopez-Peinado, A.; Rivera-Utrilla, J.; Fernandez-Morales, I.; Lopez-Garzon, F.J. *Fuel* **1987,** *66,* 113-118.
28. (Ganga Devi, T.; Kannan, M.P., University of Calicut, Kerala, India, personal communication 1991.)
29. Lowary, T.L.; Richards, G.N. *J. Wood Chem. Tech.* **1988,** *8,* 393-412.
30. Richards, G.N. *J. Anal. Appl. Pyrol.* **1987,** *10,* 251-255.
31. Halpern, Y.; Riffer, R.; Broido, A. *J. Org. Chem.* **1973,** *38,* 204-209.
32. (Edye, L.A.; Ganga Devi, T.; Kannan, M.P.; Richards, G.N., Wood Chemistry Laboaotry, unpublished results.)
33. DeGroot, W.F.; Pan, W.-P.; Rahman, M.D.; Richards G.N. *J. Anal. Appl. Pyrol.* **1988)** *13,* 221-231.
34. Shafizadeh, F.; Chin, P.P.S. *Carbohydr. Res.* **1977,** *71,* 169-191.
35. Iiyama, K.; Wallis, A.F.A. *Wood Sci. Tech.* **1988,** *22,* 271-280.

RECEIVED April 6, 1992

POLYMERS, PLASTICS, AND TIRES

Chapter 9

Cofiring Tire-Derived Fuel and Coal for Energy Recovery

D. J. Stopek[1] and A. L. Justice[2]

[1]Research & Development, Illinois Power Company, Decatur, IL 62525
[2]Used Tire Recovery Program, Office of Recycling and Waste Reduction,
Illinois Department of Energy and Natural Resources,
Springfield, IL 62704

This chapter summarizes testing conducted on utility and industrial boilers on the cofiring of tire-derived fuel (TDF) and coal. The disposal of waste tires is a growing problem of national magnitude. Industry is responding to assist government by using waste tires as a fuel supplement with coal. Not all boilers are equally suited to this task. There are risks to the plant that must be evaluated before commercial use of tire-derived fuel can become a regular feed stock to utility and industrial boilers. A summary of test results reported by industry indicate that cofiring of TDF may be a practical solution to the waste tire disposal problem.

National interest concerning removing tires from the waste stream have focused on two fronts: Tires pose a serious health threat if left in open piles; two, burial of tires is a waste of valuable landfill space. It has been estimated that there is a waste tire generated for each person in the nation, each year. It has been estimated that there are over two-billion waste tires stockpiled nationally (1). As vehicle owners replace their old tires, only about twenty five percent are recycled (2). This leaves about seventy five percent of the used tires as new waste. Industry is becoming more aware of the need to assist with this disposal problem. Industrial and utility boilers may be used as a means of disposing of large quantities of tire waste. This is not as simple as it may sound. Not all boilers are suited to firing tires. This paper describes the different types of boilers that have been used for cofiring tires with coal and summarizes data reported from these tests.

Commercial Aspects

The use of tire-derived fuel (TDF) on a wide scale is hampered by the lack of commercial shredding facilities capable of producing the material at a competitive price. This is typical of the waste recycling industry in general. The chicken-and-egg problem has been a deterrent to a variety of recycling efforts. To break this cycle in Illinois, the Department of Energy and Natural Resources (ENR) has sought to assist industry by supporting their test programs with the partial payment for costs associated with the tests *(3)*. Both Monsanto and Illinois Power (IP) were requested by the State of Illinois Department of Energy and Natural Resources (ENR) to assist with the disposal of tires in the state. To facilitate this effort, a test program was initiated to determine the viability of cofiring chopped tires, TDF, with coal on a large cyclone fired boiler at IP *(4,5)* and a stoker boiler at Monsanto *(6)*. This program has demonstrated that IP has the capability of disposing of up to six million tires a year mixed at two percent by weight with the coal at it's Baldwin Station, and that Monsanto can dispose of about one and one-half million tires, at a 20 percent by weight blend rate, at its Krummrich Plant in Sauget, Illinois.

The used tire industry has been hesitant to invest in quality tire shredding equipment on a wide scale, until a fuel market is well established. Mr. Tryggve Bakkum, of Waste Management of North America, reported that the cost for a single tire shredding machine with a throughput of about 10 tons per hour is a half-million dollars *(7)*. This industry is still developing, and striving to improve the durability of the hardware available. He estimated that tipping fees of 70 to 80 dollars per ton are needed for TDF to be widely available. Fuel cost may limit income on sales to 15 to 25 dollars per ton. Since current disposal costs are about 50 to 60 dollars per ton nationally, there is little margin for commercialization, until other methods of disposal become unavailable or until alternate disposal costs increase sufficiently to cover the cost of making TDF.

The state of Wisconsin has recognized this problem, and provides an incentive or subsidy of $20/ton to facilities that burn or use waste tires *(8)*. A variety of other states are examining similar programs.

Information Sources

Companies interested in learning more about tire burning can find assistance in their quest for information. As state pollution control agencies struggle with identifying means of coping with the tire disposal problems, they are collecting considerable information. Other sources of information include:

- Scrap Tire Management Council
  1400 K Street, N.W., Washington DC, 20005, (202) 408-7781
- Electric Power Research Institute
  Generation & Storage Division, Manager of Technical Assessment
  P.O. Box 10412, Palo Alto, California, 94303, (415) 855-2445

Both of these organizations have sponsored conferences dealing with the issues of waste tire disposal. They maintain information programs for technology transfer.

Boiler Types

There are a variety of boiler types that are used for generating steam. These are the pulverized coal, the cyclone, the stoker, and the fluidized bed boilers *(9)*. Tires have been burned in cement kilns as well. Table I lists some of the facilities known to have tested combustion of TDF with coal (2).

The most predominant type of coal-fired boiler is the pulverized coal boiler. This type of boiler requires that fuel be ground to a very fine powder, typically seventy percent less than 200 mesh. The cost to pulverize tires to this size would be prohibitive as a fuel. For this reason, they are generally not considered as a candidate for use. There are two types of pulverized coal fired boilers "dry" bottom and "wet" bottom. Dry bottom boilers are generally not suited tocofiring waste tires with coal. This is because of the steeply sloped bottom used to discharge bottom ash from the boiler, does not lend itself to supporting large particles of burning materials. Also, there is no grate or other device to hold tires in the furnace while the tire chips burn.

There is a select number of pulverized boilers that have a "wet" bottom. A wet bottom means that the fire in the boiler is sufficiently hot to melt the coal ash, which runs as a liquid slag, and collects at the somewhat flat bottom of the boiler. The molten slag is drained out the bottom of the boiler into a tank of water, where the slag is quenched. Ohio Edison has tested burning whole tires in their Toronto Station, which is a 42 MW boiler of this type (10). In order to feed whole tires into the boiler, Ohio Edison developed a lockhopper system to admit the tires to the boiler. Tires are allowed to roll into the boiler about every 10 seconds, through the lockhopper. They have successfully fed up to 20 percent of the fuel to the boiler as tires.

Stoker boilers are typically found in industrial facilities and small urban utility settings. These boilers are fed chunk coal of about an inch in size. The fuel is fed onto a grate, a set of steel bars that support the fuel while it burns. Air for combustion flows up through the grate, providing cooling to the bars. There are a variety of designs used for supporting and feeding the fuel. An example of a stoker boiler is shown in Figure 1 (9). These boilers are well suited to feeding chunky material such as tire-derived fuel (TDF), because the material is about the same size as the coal fuel. About a dozen industrial locations across the country have used or tested TDF cofiring with coal (2). These include the Monsanto Krummrich Plant (6) in Sauget, Illinois and the New York State Electric & Gas (NYSEG) Jennison Station in Bainbridge, New York (11).

Another style of boiler is the cyclone fired boiler. Illinois Power conducted tests on Units 1 and 2 at their Baldwin Station which are 560 MW, B&W cyclone-fired boilers (4,5). Other utilities that have burned TDF in cyclone boilers include Wisconsin Power & Light at their Rock River Station in Beloit, Wisconsin, (12) Otter Tail Power at their Big Stone Plant in Big Stone City, South Dakota, (13) and Northern Indiana Public Service at their Michigan City Station. The Baldwin boiler has 14 cyclones, seven each on the front and rear walls. Operation of a cyclone boiler is different from either a pulverized coal boiler or a stoker boiler. A cyclone boiler burns coal that is crushed to a size of less than ¼ of an inch. The coal is fed to the cyclone burners, Figure 2 (9). Air enters the cyclones tangentially creating a cyclonic action. The coal burns at high temperatures, typically 3000 °F. The hot gas exits the cyclone into a large furnace. At these temperatures, the ash melts into a liquid slag that flows to the bottom of the furnace, where it is quenched in a pool

Table I.   Facilities That Have Tested Cofiring of Waste Tires

COMPANY NAME	City	State
Stoker Boilers		
Champion International	Bucksport	Maine
Champion International	Sartell	Minnesota
Crown Zellerbach	Port Angeles	Washington
Fort Howard Paper	Green Bay	Wisconsin
Great Southern Paper	Ceder Springs	Georgia
Inland - Rome Paper	Rome	Georgia
NYSEG	Binghampton	New York
Port Townsend Paper	Port Townsend	Washington
Rome Kraft Pulp & Paper	Rome	Georgia
Smurfit Newsprint	Newburg	Oregon
Sonoco Products	Hartsville	So. Carolina
Traverse City Light & Power	Traverse City	Michigan
United Power Associates	Elk River	Minnesota
Willamette Industries	Albany	Oregon
Pulverized Boilers		
Ohio Edison	Toronto	Ohio
Cyclone Boilers		
Illinois Power	Baldwin	Illinois
Northern Indiana Pub. Serv.	Michigan City	Indiana
Otter Tail Power Co.	Big Stone	So. Dakota
Wisconsin Power & Light	Beloit	Wisconsin
Fluidized Bed Boilers		
ADM	Decatur	Illinois
Manitowoc Public Utilities	Manitowoc	Wisconsin

(Adapted from ref. 2:   Malcom Pirnie, Inc. and other
communications.)

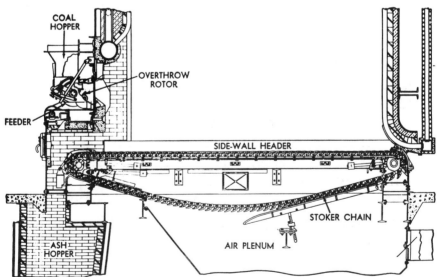

Figure 1. Typical coal feed and ash removal for a reciprocating grate (above) and a traveling grate (lower) stoker boiler. Reproduced with permission from Ref. 9, pp. 16-11 & 12, copyright 1963, B&W.

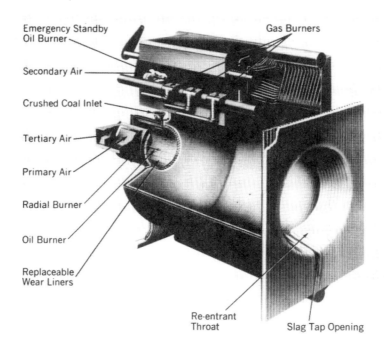

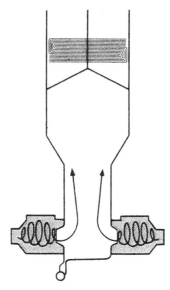

Figure 2. Arrangement of opposed firing cyclones (lower) and design of a cyclone burner (above). Reproduced with permission from Ref. 9, pp. 10-1 & 2, copyright 1972, B&W.

of water. Quenching shatters the slag and the resulting glass-like beads are a saleable product. The slag is sluiced to a pond, where it is collected by the reclaim company.

Fluidized bed boilers are becoming more commonplace over the past five years. These require fuel delivered at a size somewhat like a cyclone boiler. Several fluidized bed boilers have tested combustion of TDF with coal *(14)*. There is limited data available from these facilities, and this paper does not explore the details of their use.

Cement kilns burn coal as a source of heat and ash in the making of cement from limestone. Several cement kilns have tested the use of TDF *(2)* as a supplement or replacement of the coal as a fuel source. One advantage of tires is that the iron in the tires is a needed ingredient in the cement mix. This paper does not explore the details of the cement industry use.

Technical Concerns

The concerns of a boiler operator about to test any new fuel are increased when the fuel is a waste product, with properties that are not well known. In the case of TDF these include:

- Can tire material be burned in the boiler?
- Can TDF be delivered to the boiler reliably?
- Does TDF combustion pose hazards to the equipment?
- Will TDF affect the environmental operation of the plant?

TDF Combustion. The size of the coal fed to any boiler must be uniform and of the proper size for the style of boiler to ensure good combustion. If TDF is fed in larger pieces, there is concern that complete combustion may not occur. The sulfur content of TDF is lower than high-sulfur coal, but not low enough to meet the 1.2 lbs/MBtu compliance level required by the Clean Air Act. Other differences between TDF and coal are that the tires are high in zinc, and it is not clear whether the zinc will be collected in with the fly ash or will remain volatile. If collected, will the zinc leach out of the ash? A particular concern in a cyclone, is whether unburned material may be carried over into the dust collection equipment. Another concern is whether the TDF non-combustibles, primarily steel belting, will melt in the slag. If TDF adversely effects slag flow or consistency it can interfere with the boiler operation or byproduct sales of the slag *(15)*.

In testing conducted by Illinois Power (IP), there was no evidence of increased carbon carryover due to TDF combustion. IP tested at TDF levels of 2 percent *(5)*. Wisconsin Power and Light (WP&L) has conducted tests with TDF levels up to 10 percent *(12)*. The steel belting did not appear to create any problems. However, a small amount of unburned TDF was evident in the slag from both utilities' tests. This is a concern to IP because all of the slag is sold for grit blasting and other uses. IP plans additional testing to evaluate this aspect of the operation more thoroughly *(5)*. The TDF tested had much of the metal removed at the vendor by means of a magnetic separator. This removed most of the bead wire. There was no evidence of problems associated with the remaining tread wire.

When burning TDF in a stoker, little difficulty has been reported due to changes in the boiler operation. NYSEG noted the appearance of droplets of metal adhering to the grate *(11)*. These have the appearance of weld splatter. However, this has not been a deterrent to the operation of the unit, as the material came off with little effort. NYSEG conducted tests with TDF feed rates as high as 50 percent. Monsanto did not report any problems with their operation *(6)*.

Ohio Edison has not encountered any serious problems with their operation, using whole tires *(10)*. They have voiced some concern about the buildup of iron in the slag from the bead metal in the tires. They plan to monitor this aspect of the process closely.

Material Handling. It is important that the delivery of TDF to the boiler not interfere with the delivery of coal. The coal handling system at each boiler plant is different and unique. This requires that individual review be conducted to evaluate the best means for delivering TDF for a specific plant. However there are many common aspects to coal handling systems. Figure 3 is a diagram of a typical coal feed system for a cyclone boiler *(9)*. By and large, most facilities will prefer that the TDF be mixed with the coal near the front end of the coal handling process. Many have installed temporary conveyors to place the TDF onto the main belt feeding the station. If the TDF has poorly sized or prepared material it can greatly interfere with the coal handling equipment. Common problems include long thin ribbons of TDF becoming tangled in equipment; exposed pieces of wire locking together to form large balls that jam equipment; and loose cording collecting together to form "fur balls" that interfer with coal flow *(12)*.

IP determined that for test purposes, no plant modifications should be made. Thus, the TDF was mixed with the coal prior to the coal crushers. This means that the TDF must be capable of passing through the hammer mills, without either gumming them up or interfering with the crushing of coal or its sizing.

To facilitate coal feeding a one-yard front end loader was rented. It was used to feed coal into a hopper in the reclaim yard. TDF was fed at approximately twenty tons per hour while total material flow was fed at about 1000 tons per hour, yielding the desired mixture of two percent TDF and coal.

During testing it was determined that 2-inch TDF material could not be handled either by the crushers or the coal conveying system. This material caused over-load of the crusher motor and also plugged the coal feed chutes in the tripper (the conveyor that discharges coal to the individual storage silos above the boiler). One-inch TDF material was handled with little difficulty in the plant, although it did increase the motor amps in the crushers. If long term operation were conducted, IP would install a method to feed TDF that could bypass the crushers. Other users of TDF have been able to feed two-inch material, but most agree that it is preferable to receive one-inch TDF.

Environmental Aspects. A major area of concern is the environmental impact of TDF cofiring. Testing was conducted to determine if stack emissions were adversely effected by the TDF. These tests are also required for permit applications. For the tests conducted at IP, Table II presents the analyses of the raw fuels and the

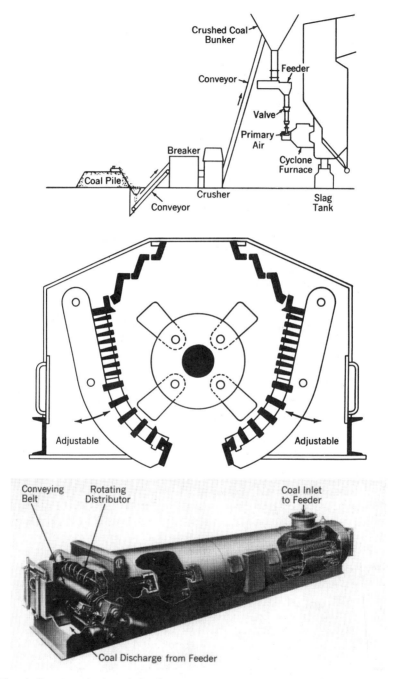

Figure 3. A typical coal feeding system (top), a hammer-mill coal crusher (center), and belt-type coal weighing feeder (bottom). Reproduced with permission from Ref 9, pp. 15-9; copyright 1963, pp. 10-2 & 4, copyright 1972, B&W.

Table II. Fuel Analysis of TDF and Coal

Fuel Analysis	Coal	TDF-A	TDF-B	TDF-C	Calc.2%
Moisture, %	12.8	1.14	0.51	.66	12.6
Ash, %	10.4	5.62	6.38	16.8	10.3
Volatile, %	34.4	68.4	62.6	68.3	35.0
Fixed Carbon, %	42.4	24.8	30.5	14.2	42.1
Btu/lb	11000	15969	15837	14129	11098
Carbon, %	61.1	80.9	80.1	69.9	61.5
Hydrogen, %	4.11	7.30	6.99	6.75	4.17
Sulfur, %	2.84	1.51	1.34	1.39	2.81
Ash Analysis, wt.%					
Silicon dioxide	60.3	23.6	18.3	4.13	59.5
Aluminum oxide	21.74	8.18	7.52	1.12	21.5
Titanium dioxide	0.70	5.60	9.20	0.28	0.86
Iron oxide	2.83	10.6	35.5	80.1	3.40
Calcium oxide	4.49	4.00	3.22	1.80	4.47
Magnesium oxide	1.47	1.28	0.70	0.38	1.46
Potassium oxide	1.02	0.92	0.52	0.20	1.01
Sodium oxide	1.25	0.92	0.70	0.16	1.24
Undetermined	2.14	2.00	1.24	0.21	2.12
Zinc oxide		33.6	17.8	8.96	0.41
Zn (ppm)	188				
Trace Metals					
As	<8	14	8	26	<8
Ba	3400	170	90	45	3334
Cd	7	28	18	<4	7
Cr	28	97	120	150	30
Pb	<20	140	180	40	<20
Hg	<0.02	0.02	0.03	0.02	<0.02
Se	<8	<8	<8	<8	<8
Ag	<4	<4	<4	<4	<4

(Adapted from ref. 5.)

Table III. Range of Data, ESP Emission Test - TDF & Coal

	LOW	HIGH
COAL ONLY		
Inlet Dust Conc. lbs/hr	16,800	17,700
Outlet Dust Conc. lbs/hr	670	710
Outlet Dust Loading, lbs/MBtu	0.13	0.14
ESP Efficiency, %	95.6	96.2
COAL & TDF		
Inlet Dust Conc. lbs/hr	16,400	19,400
Outlet Dust Conc. lbs/hr	840	1,030
Outlet Dust Loading, lbs/MBtu	0.163	0.197
ESP Efficiency, %	93.8	95.7

(Adapted from ref. 5.)

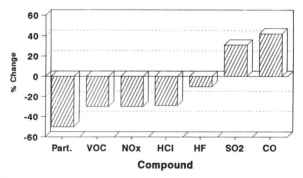

Figure 4. A comparison of stack emissions between the baseline conditions of coal only and when burning a 20% blend of TDF. Reproduced from Ref. 6.

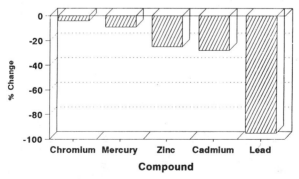

Figure 5. A comparison of metals stack emissions between the baseline conditions and when burning a 20% blend of TDF. Reproduced from Ref. 6.

calculated values for the mixed fuels. TDF analyses A and B are from two different suppliers of one-inch debeaded material. TDF C is an analysis of two-inch TDF that includes the bead material. Note that the removal of the bead material increases the heating value of the fuel corresponding to the reduction in the ash material. As stated above, testing was discontinued with TDF C because of handling problems. No discernable difference in opacity was noted as a result of TDF cofiring. Normal opacity readings of 15 to 20 percent were recorded. Control board readings were collected to perform boiler efficiency calculations. Heat rates of about 9600 Btu/KWh were maintained during the performance period. Particulate emission tests were conducted on March 20 (baseline) and on March 21, 1991, by a consultant. The range of data from these tests is listed in Table III. The differences between the results for the two days are within the accuracy of the instrumentation and the variation in the coal quality for the period of operation. Examination of ESP power settings showed little variation as a result of TDF cofiring compared to normal operation with coal *(5)*.

Testing conducted by Monsanto on their stoker boiler revealed that cofiring TDF with coal resulted in substantial reductions of emissions *(6)*. This is shown in Figures 4 and 5. Similar results have been reported by WP&L and Ohio Edison *(12,10)*. NYSEG experienced some reduction in particulates with TDF at both 25 and 50 percent blends with coal *(11)*.

Unburned Material. One aspect of the test program was identified as a cause for further investigation by IP. Several dozen pieces of substantially unburned TDF were found in the boiler slag. Due to the nature of slag discharge, it is very difficult to determine a quantitative amount of this material. Since IP sells all of the slag from Baldwin Units 1 and 2 to a local company, the presence of the TDF remains an area of concern. Additional testing will be required to evaluate the possible cause. The results of these future tests will determine the requirements for plant modifications, if any needed to begin commercial TDF cofiring *(5)*.

WP&L has had similar experiences and has screened their slag before selling it as aggregate.

Conclusions

With the assistance of state agencies, utilities and industrial boiler operators will continue to test cofiring TDF with coal. As plant operators learn to cope with this new fuel, the chicken-and-egg supply problem will end. As financing for facilities becomes more readily available, more operations that can produce TDF at competitive fuel prices will be realized. This will allow our country's industry to help solve a national waste disposal problem.

References

1.   Illinois Department of Energy and Natural Resources. Office of Solid Waste and Renewable Resources. Illinois Scrap Tire Management Study. Springfield, IL.: October, 1989. ILENR/RR-89/04

2.   Air Emissions Associated with the Combustion of Scrap Tires for Energy Recovery. Malcom Pirnie, Inc. for the Ohio Air Quality Development Authority. May 1991.
3.   Illinois Waste Tire Management Program. Justice, A. and Purseglove, P., Illinois ENR and EPA. Scrap Tire Management Council Conference. September 1991.
4.   Testing of Tire-Derived Fuel at Baldwin Unit 1; 560 MW Cyclone Boiler, Phase I and II. Illinois Power Co. ENR Grant NO. SWMD23, Jan. 1991.
5.   Testing of Tire-Derived Fuel at Baldwin Unit 1; 560 MW Cyclone Boiler. Illinois Power Co. ENR Grant NO. SWMD25, November, 1991.
6.   Test Burning of Tire-Derived Fuel in Solid Fuel Combustors. Dennis, D., Monsanto Co. for Ill. ENR. ILENR/RR-91/16(ES), July 1991.
7.   Tire Shredding Equipment. Bakkum, T. and Felker, M., Waste Management of North America. Proceedings of 1991 Conference on Waste Tires as a Utility Fuel, EPRI-GS-7538, September 1991.
8.   Overview of Regional Waste Tire Management Opportunities for Electric Utilities. Koziar, P., Wisconsin Department of Natural Resources. Proceedings of 1991 Conference on Waste Tires as a Utility Fuel, EPRI-GS-7538, September 1991.
9.   Steam, Its Generation and Use. Babcock and Wilcox Co. 1963, Thirtieth Edition and 1972, Thirty-Eighth Edition.
10. Ohio Edison Tire Burning Project. Gillen, J., Ohio Edison. Pittsburgh Coal Conference, October 1991.
11. Tires to Energy, Research Project Report, Tesla, Michael R., New York State Electric and Gas Corp. September 1991.
12. Experience with Tire-Derived Fuel in a Cyclone-Fired Boiler. Hutchinson, W., Eirschele, G., and Newell, R., Wisconsin Power and Light Co. Proceedings of 1991 Conference on Waste Tires as a Utility Fuel, EPRI-GS-7538, September 1991.
13. Tire Derived Fuel and Lignite Co-Firing Test in a Cyclone-Fired Utility Boiler. Schreurs, Stuart T., Otter Tail Power Co. Proceedings of 1991 Conference on Waste Tires as a Utility Fuel, EPRI-GS-7538, September 1991.
14. Manitowoc Coal/Tire Chip - Cofired Circulating Fluidized Bed Combustion Project. Phalen, J., and Taylor, T.E., Foster Wheeler Energy Corp., and Libal, A.S., Manitowoc Public Utilities. Proceedings of 1991 Conference on Waste Tires as a Utility Fuel, EPRI-GS-7538, September 1991.
15. Fuel Characterization of Coal/Shredded Tire Blends. Granger, John E. and Clark, Gregory A., B&W Co. Proceedings of 1991 Conference on Waste Tires as a Utility Fuel, EPRI-GS-7538, September 1991.

RECEIVED June 22, 1992

Chapter 10

Converting Waste Polymers to Energy Products

P. Assawaweroonhakarn and J. L. Kuester

Department of Chemical, Biological, and Materials Engineering, Arizona State University, Tempe, AZ 85287–6006

A project was performed to determine the feasibility of converting waste polymers into diesel fuel. The primary waste polymer source of interest was disposable diapers, consisting of a mixture of cellulosic and synthetic polymer material in the presence of biological wastes. The overall project consisted of five phases: (1) reaction equilibrium calculations, (2) batch pyrolysis study, (3) continuous pyrolysis study, (4) continuous liquefaction study, and (5) integrated system demonstration. The integrated system consists of a circulating solid fluidized bed pyrolysis system to produce a synthesis gas for a fluidized bed catalytic liquefaction reactor. The objective for the pyrolysis system is to optimize the composition of hydrogen, carbon monoxide and ethylene in the synthesis gas for conversion to diesel fuel in the catalytic liquefaction reactor. The liquefaction reactor produces a product very similar to commercial No. 2 diesel fuel. In this chapter, results for the first two phases will be presented.

Approximately 18% by volume of municipal solid waste consists of waste plastics (1). The 15.8 billion disposable diapers used annually comprise about 2% of the total waste stream. Due to increasing landfill costs and environmental and regulatory pressure, a flurry of activity has emerged to seek alternatives to landfill disposal of these materials. Proctor & Gamble, for example, has announced an "accelerated composting" program for their disposable diaper products (2). Separation steps are implemented to segregate the noncompostable parts from the cellulosic parts. The intent is to convert the cellulosic parts into "soil enhancer" (compost). Presumably, the plastics part still goes to the landfill. The questions would be the market for the "soil enhancer" and the probable necessity to still landfill the plastics. An alternative approach would be to convert the bulk diaper (all components) into marketable products with minimal landfill requirements. This approach has been

0097–6156/93/0515–0117$06.00/0

developed at Arizona State University (ASU) utilizing over 100 different feedstocks, generally falling into the categories of industrial wastes, municipal wastes, hazardous wastes and various agricultural and forest residues. An indirect liquefaction approach is used, i.e., gasification of the feed material to a gas followed by liquefaction of the gas to a No. 2 diesel grade transportation fuel. The sequence is illustrated in Figure 1. The objective in the gasification step is to maximize the production of hydrogen, carbon monoxide and ethylene while the objective in the second step is to maximize the production of diesel fuel from these three reactants. A high octane product can be produced via conventional catalytic reforming of the diesel material. The potential products are thus liquid hydrocarbon fuels, medium quality gas (ca. 500 BTU/SCF) and/or electricity (via heat recovery or combustion of the fuels). Alternative operating conditions and catalysts for the second stage reactor could produce other products (e.g., alcohols, methane etc.).

Prior work on this process has been described elsewhere (see, for example, references 3-6). This paper will present the application for disposable diapers. A five phase approach was used (reaction equilibrium calculations, batch pyrolysis study, continuous pyrolysis study, continuous liquefaction study, integrated system demonstration) with performance limited to an 11 month period. Results for the first two phases will be presented here. The intent was to minimize the project risks and costs for scale-up to a commercial configuration. Using the composition of a particular disposable diaper product ("Huggies") as an example, the maximum yields of diesel fuel (wet and dry basis) are illustrated in Figure 2. Realistic actual liquid product yields are expected to be in the 50-100 gals/ton range (dry basis).

Equilibrium Calculations

Composition analysis for three commercial disposable diaper products is shown in Table I. Since the compositions are similar, Huggies will be used for example calculations. Assuming a gas product slate of hydrogen, carbon monoxide, ethylene, ethane, acetylene, methane and carbon dioxide, the equilibrium product composition was calculated as a function of temperature by minimizing the Gibb's free energy of formation subject to atom balance constraints. The samples under study are composed of carbon, hydrogen and oxygen as major components. The atomic ratio of Huggies for both wet and dry cases is shown in Table II. The computer code utilized to calculate the equilibrium composition used the Gibb's free energy of formation for each expected product to calculate a set of primary reactions with the associated equilibrium constants and the equilibrium mole fraction of each compound in the system. The program used the technique of Myers (7) to estimate an initial composition. A modified Newton-Raphson technique was used in solving the problem. All calculations were performed using an IBM personal computer Model XT with 8087 math co-processor. The thermodynamic data of the expected products is given in Table III (8,9).

The calculated synthesis gas compositions at equilibrium as a function of temperature for both wet and dry samples are shown in Figures 3-6. As shown, all components decrease with temperature except hydrogen and carbon monoxide. Although the dry samples gave a higher hydrogen/carbon monoxide mole ratio, the wet sample results are of more interest since the real processing feedstock is in a wet condition. The desired hydrogen/carbon monoxide ratio of ca. 1.2 (based on prior work) is achieved at ca. 1100 K

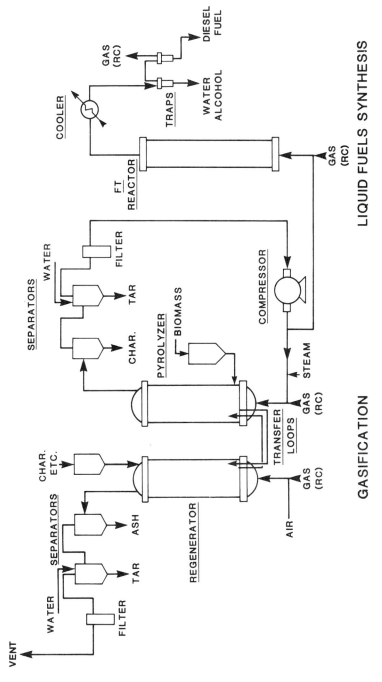

Figure 1. Conversion System Schematic.

HUGGIES ————————————→ DIESEL

$C_{2.3}H_{15.7}O_{7.1}$ ————————→ $C_{10}H_{22}$ + Other

	Composition (wt %)		
	Huggies (wet)	Huggies (dry)	Diesel
C	17.54	53.4	84.51
H	9.9	7.6	15.49
O	71.92	37.19	0
N	0.0	0.0	0
Na	0.5	1.70	0

$$\frac{\text{maximum yield}}{\text{(carbon balance)}} : \quad 67 \ \frac{\text{gals diesel}}{\text{ton Huggies (wet)}}$$

$$189 \ \frac{\text{gals diesel}}{\text{ton Huggies (dry)}}$$

Figure 2. Diaper Chemistry and Maximum Product Yields.

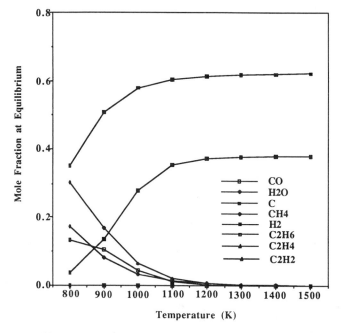

Figure 3. Equilibrium Gas Compositions of Huggies (dry) as a Function of Temperature.

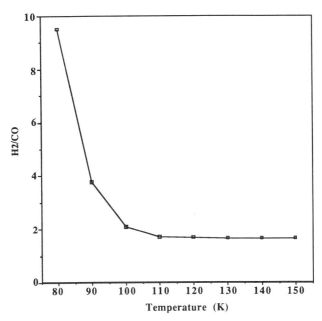

Figure 4. Equilibrium Hydrogen to Carbon Monoxide Mole Ratio of Huggies (dry) as a Function of Temperature.

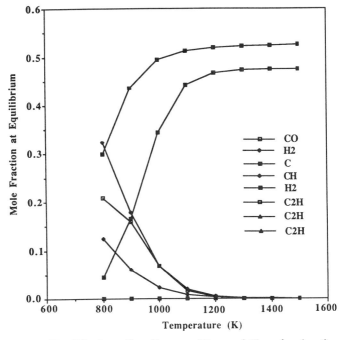

Figure 5. Equilibrium Gas Compositions of Huggies (wet) as a Function of Temperature.

Table I. Elemental Analysis of the Waste Polymer Samples

Compound	Element	Sample 1 Huggies	(gmole) Sample 2 Luvs	Sample 3 Pampers
Cellulose	carbon	1.3	1.3207	1.4417
	oxygen	1.09	1.1073	1.2088
	hydrogen	2.17	2.2045	2.4065
Polypropylene	carbon	0.521	0.2402	0.1865
	hydrogen	1.04	0.4794	0.3723
SAM	carbon	0.153	0.2142	0.2039
	hydrogen	0.167	0.2338	0.2226
	oxygen	0.102	0.1428	0.136
	sodium	0.038	0.0532	0.0507
Polyethylene	carbon	0.2	0.3288	0.304
	hydrogen	0.4	0.6576	0.608
Styrenes	carbon	0.108	0.1635	0.3742
	hydrogen	0.143	0.2165	0.4955
Polyurethane	carbon	0.0126	0	0
	hydrogen	0.0168	0	0
	oxygen	0.0056	0	0
	nitrogen	0.0014	0	0
Urine	carbon	10.72	10.72	10.72
	oxygen	5.36	5.36	5.36
Feces	carbon	1	1	1
	oxygen	0.5	0.5	0.5
Total	carbon	2.2946	2.2674	2.5104
	hydrogen	15.6568	15.5118	15.825
	oxygen	7.0576	7.1101	7.2048
	nitrogen	0.0014	0	0
	sodium	0.038	0.0532	0.0507

Unaccounted for grams			
Urine	3.6	3.6	3.6
Feces	3	3	3
Rubber	0.33	0.075	0.075

Source: Kimberly-Clark Corporation.

Table II. Atomic Ratio of Samples in Task 2

	Huggies	
Element	*Dry*	*Wet*
Carbon	1	1
Oxygen	0.5219	3.0757
Hydrogen	1.7157	6.8233

(1550 F) for the wet sample. No ethylene is predicted at equilibrium, as expected in the presence of hydrogen.

Batch Pyrolysis Study

These experiments were performed in a Chemical Data Systems Model 122 Pyroprobe coupled to a Carle Gas Chromatograph Model AGC111H . The Pyroprobe consists of a temperature programmed ampule containing a sample of the desired feedstock with product gas being swept to the gas chromatograph via helium carrier gas. For hydrogen determination, a hydrogen transfer tube was used at a temperature at about 600 C in a nitrogen stream. A SpectraPhysics Model 4270 integrator was used for the quantitative gas analysis. An IBM 9000 computer was connected to the integrator via an RS-232 interface. The computer was operated in a terminal mode using IBM 3101 terminal emulation software. The computer was mainly used to upload or download files to the disc drive. An additional CIT-101 terminal was also used for convenient communication between the operator and the integrator. The overall experimental system is schematically represented in Figure 7.

The feedstocks are a mixture of synthetic and natural polymers. Approximately 3 grams of each dry and clean sample was prepared by using the information from Table I but without the biologically hazardous material. The composition of the sample in the 3 grams mixture is shown below.

Sample 1: Huggies

Bag #1:	Polyethylene, Polypropylene, and others	0.6999 g
Bag #2:	Fluff cellulose and SAM	2.1131 g
Bag #3:	Tissue cellulose	0.1870 g

All of these components were received from Kimberly-Clark Corporation. Polymers and tissue cellulose were delivered in approximately 1 cm. size. After the size reduction process to 60 mesh had been accomplished by a means of Thomas-Wiley Intermediate Mill Model 3383, each component was weighed and well mixed in a PICA Blender Mill Model 2601 and cut again using the intermediate mill. The experimental procedure was as follows:

1) Use a quartz capillary tube, ca. 0.2 cm. diameter and 2.5 cm. long as a sample holder. Plug a small piece of quartz wool at one end of the tube. Load and weigh ca. 0.0008 g of a sample in the capillary tube, then plug the other end of the tube with a small piece of quartz wool. The sketch of a sample cell is shown in Figure 8.

2) Place the tube in the coil probe smoothly until the end of the thermocouple is imbedded in the ampule as shown in Figure 8.

Table III. Gibb's Free Energy of Formation (kcal/K) at Temperature (K)

Species	800	900	1000	1100	1200	1300	1400	1500
C	0	0	0	0	0	0	0	0
CH_2O	-22.749	-21.898	-21.024	-20.134	-19.228	-18.314	-17.391	-16.461
CH_4	-0.533	2.029	4.625	7.247	9.887	12.535	15.195	17.859
CO	-43.612	-45.744	-47.859	-49.962	-52.049	-54.126	-56.189	-58.241
CO_2	-94.556	-94.596	-94.628	-94.658	-94.681	-94.701	-94.716	-94.728
C_2H_2	43.137	41.821	40.522	39.234	37.96	36.69	35.432	34.177
C_2H_4	24.628	26.514	28.431	30.373	32.334	34.302	38.266	40.246
C_2H_6	16.010	21.11	26.26	31.4	36.61	41.8	47.02	52.2
C_3O_2	-33.224	-34.653	-36.075	-37.495	-38.906	-40.315	-41.714	-43.11
HCO	-12.307	-13.378	-14.428	-15.462	-16.479	-17.484	-18.475	-19.456
H_2	0	0	0	0	0	0	0	0
H_2O	-48.646	-47.352	-46.04	-44.712	-43.371	-42.022	-40.663	-39.297
H_2O_2	-12.181	-9.519	-6.85	-4.173	-1.495	1.186	3.87	6.554
O_2	0	0	0	0	0	0	0	0

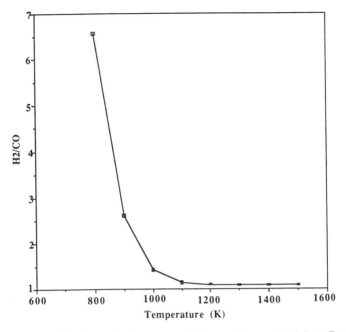

Figure 6. Equilibrium Hydrogen to Carbon Monoxide Mole Ratio of Huggies (wet) as a Function of Temperature.

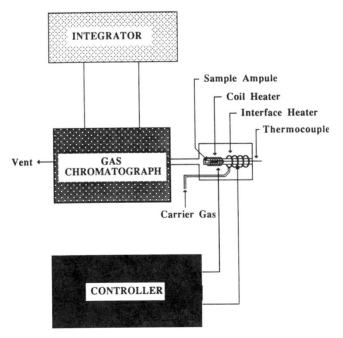

Figure 7.   Experimental Apparatus for Small Scale Batch Reactor.

3) Place the probe in the interface, maintained at ambient temperature. Set time (20 seconds) and desired final temperature on the controller.
4) Wait at least 15 minutes to allow the system to equilibrate by means of helium passing through cell.
5) Start the integrator and the temperature recorder. The gas chromatograph is started automatically by the integrator.
6) When the gas chromatograph goes to the injection mode, push the pyroprobe start button.
7) Record the highest temperature (reaction temperature) from the temperature recorder.
8) After this step, all data gathering, gas composition calculations, and experimental termination are automatic.
9) After the completed gas chromatograph cycle, removed the probe and cap the interface.
10) Remove the capillary tube from the probe.

Safety precautions and procedures for this apparatus consisted of the following:
1) Wear a dust mask when preparing samples.
2) Study the Material Data Safety Sheets for each chemical before it is used or produced.
3) Keep all drinks and flammable liquids away from the pyroprobe/gas chromatograph area.
4) Turn on the carbon monoxide emergency detector to sense any gas that may be in the area.
5) Review the equipment manuals and Laboratory Safety Manual for further information.

A full factorial designed experiment was performed for the factors temperature and water composition. The experimental design is given in Figure 9 with the results shown in Table IV using Huggies as the example feedstock. The base condition for temperature was set by results from the reaction equilibrium calculations. The three responses of interest are: (1) hydrogen + ethylene + carbon monoxide, (2) hydrogen/carbon monoxide, and (3) ethylene. The superior level for all three responses is at the high level for each factor (experiment E). The factor ranking (via analysis of variance calculations) for each response and experimental error (as calculated by base point replication range divided by factorial experiment range) is shown in Table V. The error could be caused by several factors such as heterogeneity of sample and the platinum coil condition. All response results are considered to be favorable for further investigation in a continuous system (phases 3, 4 and 5 of the project).

A complete mass balance was not possible for this phase due to the small sample size (0.0008 grams). Thus the distribution of polymer to gas, char and tar, was not reported for this batch reactor phase.

Summary and Conclusions

The production of liquid hydrocarbon fuels from biologically contaminated disposable diapers is a technically viable concept. Favorable synthesis gas compositions can be produced. High quality liquid hydrocarbon fuel products are expected to be produced from the synthesis gas. Additional research

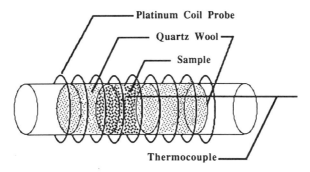

Figure 8.   Sample Cell and the Coil Probe.

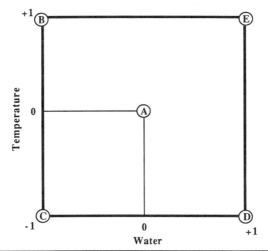

	Level		
Factor	*-1*	*0*	*+1*
Temperature (K)	1019	1102	1185
(°F)	1375	1525	1675
Water (wt%)	0	64.3	78.3
(μ 1/0.1 mg.)	0	0.18	0.36
Pressure:		1 atm	
Residence Time:		20 seconds	

Figure 9.   Experimental Design for Huggies.

128   CLEAN ENERGY FROM WASTE AND COAL

Table IV. Gas Compositions from Huggies (Factors: Temperature and Water Addition)

Experiment	A	B	C	A	D	E
◊ Operating Condition						
Coil Temperature (K)	1373	1467	1277	1373	1277	1467
(°F)	2012	2181	1839	2012	1839	2181
Reaction Temperature (K)	1093	1176	1020	1108	1006	1172
(°F)	1508	1658	1376	1535	1352	1651
Water Added (wt %)	64.3	0	0	64.3	78.3	78.3
(μ l/0.1 mg of sample)	0.18	0	0	0.18	0.36	0.36
◊ Gas Composition (Mole %)						
H_2	26	29	14.5	28	22	32
CO_2	19	19	36	16	20	12
C_2H_4	7	5	5	7	6	7
C_2H_6	2	1	2	2	3	1
C_2H_2	0	0.2	0	0	0	2
CH_4	13	11	10.8	14	14	14
CO	33	34	30.7	33	35	32
H_2/CO Ratio	0.80	0.83	0.47	0.85	0.65	0.98

Table V. Analysis of Variance Calculation Results

Response	Factor Rank	% Error
$H_2 + CO + C_2H_4$	Temperature, Water	9.6
H_2/CO	Temperature, Water	9.8
C_2H_4	Water, Temperature	0

and development work is required to establish reliable mass and energy balances for the continuous system before scale-up is considered. To accomplish this objective, a reliable solids feeder system needs to be designed and tested. This is not considered to be a major obstacle.

Literature Cited

1. *Pressure for plastic recycling prompts a mix of tough laws and cooperation*; In Wall Street Journal, February 2, 1990, pp. B7.
2. Procter & Gamble advertisement. *This baby is growing up in disposable diapers*; In Newsweek, October 15, 1990.
3. Kuester, J.L. *Bioresource Technology*. **1991**, *35*, 217-222.
4. Kuester, J.L. *Electrical power and fuels production from wastes*; In Instrumentation in the Power Industry; ISA: 1989; Vol. 32, pp. 69-74.
5. Prasad, B.; Kuester, J.L. *Industrial and Engineering Chemistry Research*. **1988**, *27*, 304-310.
6. Davis, E.,; Kuester, J.L.; Bagby, M. *Nature*, **1984**, *307*, 726-728.
7. Myers, A.; Myers, A. *J. Chem. Phys.* **1986**, *84*, 5787-5795.
8. Dow Chemical Company. *JANAF Thermochemical Tables*. U.S. Bureau of Standards: Washington, D.C.,1971.
9. Chao, J.; Wilhoit, R.C.; Zwolinski, B.J. *J. Phys. Chem. Ref. Data.* **1973**, *2*, 427-437.

RECEIVED April 6, 1992

Chapter 11

Gasification–Pyrolysis of Waste Plastics for the Production of Fuel-Grade Gas

L. L. Sharp and R. O. Ness, Jr.

Energy and Environmental Research Center, University of North Dakota, Box 8213, University Station, Grand Forks, ND 58202

New technologies in gasification, pyrolysis, and environmental control are allowing the use of waste materials for feedstocks in environmentally acceptable ways for the production of electricity or process heat. Ebonite and automotive shredder residue were tested in a TGA using potassium- and calcium-based catalysts, and ebonite was tested in a 1- to 4-lb/hr continuous fluid-bed reactor (CFBR). Analysis of the data determined that fuel-grade gases could be produced in short residence time periods. Gas compositions of over 50 percent hydrogen were recorded with conversions of over 90 percent. Data include carbon conversion, gas production, wastewater treatment, heavy metals analysis, and chlorine content.

Many areas of the United States are experiencing waste disposal problems stemming from a shortage of landfill space, a public concern for the environmental impacts, and the appearance of landfilling. Incineration, another popular waste disposal method, is facing more stringent emissions regulations, has a poor environmental record, and has no additional benefits aside from the disposal of the waste. Although landfilling is still the cheapest method of disposal, the cost is rising due to the implementation of environmentally acceptable landfilling techniques. Thus cost-competitive alternative methods of recycling and conversion of waste to other products such as fuel are becoming more feasible. Table I lists some alternative processes which have been proposed and studied over the past several years.

New gasification and pollution control technologies are making it possible for waste materials to be used as environmentally acceptable sources of energy, providing both disposal and energy production. Two possible waste materials which can be used as feedstocks for energy production: automotive shredder

0097–6156/93/0515–0129$06.00/0

Table I. Summary of Proposed Processes

Company/Process	Description
Argonne National Labs	Separation and recycling of plastic components of ASR. Uses solvent extraction to separate out plastics.
Voest-Alpine	High-temperature gasification (3000°F).
Cookson/deTOX	Ash treatment. Ash is subjected to submerged arc melting, producing vitrified product.
Puremet	Ash treatment. Removes nonferrous metals from ASR using a cuprous ammonium sulfate solution.
Energy Products of Idaho	Fluidized-bed combustion.
RETECH, Inc./Plasma Centrifugal Reactor	Stabilization of waste materials by vitrifying the solid components in a plasma furnace.
EnerGroup, Inc.	Rotary kiln combustion.

residue (ASR) and ebonite, are being investigated at the Energy and Environmental Research Center (EERC) at the University of North Dakota (UND). This investigation includes some of the special concerns that arise from the use of waste materials in gasification, such as the fate of heavy metals, and the general suitability of these waste materials as gasification feedstocks.

Automotive Shredder Residue (ASR)

Production of Automotive Shredder Residue. Every year eight to ten million cars and trucks are disposed of by shredding at one of the 200 auto shredders located in the United States. ASR is a waste product generated in the dismantling of automobiles by the following procedure, illustrated in Figure 1. An automobile is stripped of its gas tank, battery, tires, and radiator. It is beneficial to have these items removed for safety and environmental concerns, but this is not always accomplished. After removal of some or all of these items, the automobile is shredded to provide a material less than four inches in size and composed of approximately 50 percent organic and 50 percent inorganic fractions. Magnetic separation is then used to sort out ferrous scrap. Twelve to 14 million tons of scrap per year are supplied to the steel industry for electric arc furnace feedstock from the dismantling of automobiles. Air cyclone separators isolate a low density "fluff" from the nonferrous fraction (aluminum, copper, etc.). This fluff (shredder residue) is composed of a variety of plastics, fabrics, foams, glass, rubber, and an assortment of contaminants (*1*).

The bulk density of the fluff is approximately 20 lb/ft^3. A typical composition of fluff is shown in Table II.

In the U.S., approximately 2.5 to 3 million tons per year of shredder residue are generated in this process (*1*). This is equivalent to filling 500-700 football fields ten feet deep with ASR each year. This figure is expected to rise, since cars are increasingly manufactured with more and more plastics. One study estimates that the fraction of ASR will double between 1987 and 1997, based on the relative amounts of plastics in 1977 and 1987 model cars and the fact that the average age of a car being shredded is 10 years. The cost of landfilling all of this material is now between $12 to $100 per ton, depending on location (shipping not included) (*1*).

Along with autos, "white goods," or old appliances such as refrigerators and washing machines, are disposed of in combination with the automobiles in the shredders. White goods are the main contributor of polychlorinated biphenyls (PCBs) in ASR. Shredder residue also contains a wide variety of heavy metals and halogens, making it a good candidate test feed material for gasification, as it will present many of the common problems to be considered when using waste as a gasification feedstock.

Several additional difficulties need to be addressed when dealing with ASR. The most troublesome are the heavy metals (especially cadmium) content and the previously mentioned PCBs. Heavy metals, in addition to cadmium, that need to be addressed include lead, arsenic, barium, chromium, selenium, and mercury. Table III is a typical analysis from a shredder facility (analysis prepared by Analytical Industrial Research Laboratory, Chattanooga, TN, Lab No. 90356-303, 1990). Lead in ASR comes from items such as car batteries, wheel weights, exhaust systems, body repair filler, and highway contaminants. Lead in ash resulting from processing automotive shredder residue occurs mainly in the form of lead chloride, which indicates a high leachability (*2*). Shredder residue is not yet considered a hazardous waste by the United States Environmental Protection Agency, but could conceivably be classified as such in the future. Fluff has been classified hazardous in the state of California because of its cadmium content (*1*). Any process selected to treat shredder residue will have either to eliminate heavy metals before processing, for instance with leaching, or be able to deal with these substances in the ash and/or wastewater. Another problem associated with fluff as a feedstock is the feed variability. ASR provides a wide range of feedstock composition, as shown in Table IV. A process that can gasify this feedstock should be able to handle most plastic feedstocks.

An additional consideration when choosing automotive shredder residue as a test material in the study of waste material gasification is the urgency associated with the problem. In three to four years, the rising cost of landfilling will make shredder operations unprofitable. In order to offset the cost of ash stabilization and disposal, the volume of fluff needs to be reduced. Additional revenues can be generated by the production of electricity or fuel. Due to the fact that fluff has a relatively medium heating value, 5400 Btu/lb, any process for volume reduction which takes advantage of this energy-producing potential seems a logical course of action.

Table II. Typical Auto Fluff Composition

	Dirt, Stone, and Glass Fines Removed (%)	After Screening and Trommeling (%)
Fiber	42.0	47.8
Fabric	3.1	3.6
Paper	6.4	7.3
Glass	3.5	0.5
Wood	2.2	2.5
Metals	8.1	0.5
Foam	2.2	2.5
Plastics	19.3	22.0
Tar	5.8	6.6
Wiring	2.1	0.5
Elastomers	5.3	6.2
	HHV = 5400 Btu/lb	HHV = 6163 Btu/lb

Bulk Density: Approx. 20 lb/ft^3

SOURCE: Reproduced from ref. 2.

Table III. Automotive Shredder Residue

pH	6.4
PCBs	19 ppm
Odor	Oil
Color	Black and Brown
Phenolics	8.25 ppm
Cyanides	< 0.031 ppm
Sulfides	2.7 ppm
Flash Point	> 140°F
Physical State	Solid
% Free Liquids	None
Specific Gravity	0.0432

EP Toxicity Metals

Arsenic	< 0.002 ppm	Lead	0.81 ppm
Barium	< 0.1 ppm	Mercury	< 0.0002 ppm
Cadmium	0.92 ppm	Selenium	< 0.002 ppm
Chromium	< 0.05 ppm	Silver	< 0.01 ppm

Used with permission

Table IV. Shredder Fluff Proximate/Fuel Analysis

	High	Low	Average
H_2O, wt%	34	2	10
Ash, wt%	72	25	44
Volatiles, wt%	66	24	44
Fixed carbon, wt%	12	0	3
Sulfur	0.5	0.2	0.4
Chlorine	16.9	0.7	3.4
HHV, Btu/lb	9,260	2,900	5,400
maf HHV, Btu/lb	12,830	9,930	11,600

SOURCE: Adapted from ref. 2.

Shredder operators currently raise operational revenue by selling scrap metal to the steel industry for use in electric arc furnaces. It is highly desirable to recycle scrap metal from a steel manufacturer's point of view. When an arc furnace is charged with scrap, instead of iron ore, a 74 percent energy savings is realized (2). Shredders, then, not only dispose of a tremendous number of unwanted cars and trucks, but provide the steel industry with a valuable feedstock. Shredder residue is a problem that must be dealt with soon if the shredder industry is to continue to be a viable member of the recycling community.

Automotive Shredder Residue Gasification/Pyrolysis Results

TGA Procedure.   Tests with ASR were conducted in the thermogravimetric analysis (TGA) instrument to determine test matrix conditions for further experiments to be run in the 1- to 4-lb/hr CFBR (continuous fluid-bed reactor). Two tests were conducted to determine reactivity with steam and one catalyst at temperatures of 800° and 900°C.

The TGA graphs in Figures 2 and 3 start at time = 0 (Point A). This point marks the beginning of the heatup period. Weight loss during the heatup period (as measured by a decreasing weight percent value) is due to moisture loss and devolatilization. Devolatilization, determined by prior proximate analyses, is completed before the introduction of steam (Point B). When reaction temperature is reached, steam is added, and the temperature is held at a constant value. When 50 percent of the fixed carbon is converted (Point C), heat is turned off and the steam flow terminated. The point at which 50 percent of the fixed carbon is converted is also determined by examination of proximate analysis data. As stated, each reactivity test was terminated when about half of the fixed carbon of each devolatilized sample had been converted to gas: this point is at about 22 weight percent (1 percent fixed carbon and 21

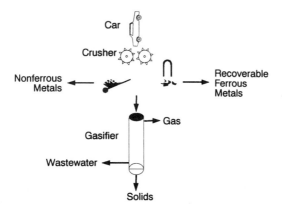

Figure 1. Automotive Shredder Residue Flowchart

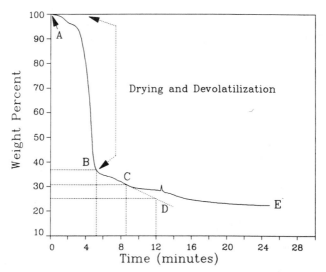

Figure 2. Automotive Shredder Residue at 900°C - No Catalyst

percent ash) of the original sample in the case with catalyst and about 28 weight percent (2 percent fixed carbon and 26 percent ash) of the original sample in the case with no catalyst. Theoretical 100% carbon conversion is shown at Point D. After the termination of heat and steam, the conversion line still continued until the instrument was turned off (Point E), since some residual steam was still present. All tests were performed under flowing argon gas.

To determine the point at which half of the fixed carbon had been converted, TGA proximate analyses were carried out on each sample prior to the reactivity tests. These analyses, summarized in Table V, showed that volatiles are removed from the reactant mixtures, and the remaining material consists of about 2 percent fixed carbon and 21 percent ash for ASR with catalyst and about 4 percent fixed carbon and 26 percent ash for ASR without catalyst. The increased ash content in the uncatalyzed sample was due to variation in the amount of metals within the original sample. Size of the ASR was -1/8 inch in all tests.

TGA ASR Results. In the case of ASR without catalyst at 900°C, time for complete conversion was approximately seven minutes (Figure 2). Conversion is calculated by subtracting the 100% conversion point (Point D) from the point where conversion began (Point B). A similar curve is also shown for ASR with K_2CO_3 as a catalyst (Figure 3). Time for conversion for ASR with catalyst at 800°C was also seven minutes.

Reactivities at both of these conditions are shown in Table VI. Reactivity changes were slight and only appeared to affect the gasification of the char, which amounts to only a small portion of the entire feed samples, so that no significant variation appeared in the two TGA tests. Gas analyses from the two runs are shown in Table VII.

During continuous fluid-bed tests, devolatilization and gasification would occur simultaneously and would be dominated by devolatilization. Since a bed material will be necessary to prevent entrainment and for sulfur capture, the ASR should have sufficient residence time to devolatilize completely and gasify. Since devolatilization was the primary reaction, gasification may not be justifiable because of the small amount of fixed carbon remaining.

Ebonite

Production of Ebonite. Ebonite is a hard rubber material used primarily in car battery casings. Like automotive shredder residue, ebonite contains a variety of heavy metals and halogens, making it a suitable test material for waste gasification tests. Ebonite is a more homogeneous feedstock than ASR and thus will have fewer processing difficulties. Specifically, since ebonite is similar in density to coal, for which the EERC system was designed, it proved easier to feed to the reactor and fluidized quite well. Proximate and ultimate analyses for the ebonite are shown in Table VIII.

Table V. Automotive Shredder Residue - Proximate Analysis

	As Rec'd	mf	With K_2CO_3 As Rec'd	mf
Moisture, wt%	3.92	-----	2.85	-----
Volatiles, wt%	67.47	70.22	74.64	76.83
Fixed carbon, wt%	3.93	4.09	2.18	2.24
Ash, wt%	24.71	25.72	20.35	20.95

Table VI. ASR Carbon-Steam Reactivities

Run Temperature, °C	Temp. of Steam Addition, °C	Reactivity Constant, k
900	900	11.80
800	800	15.70

Table VII. TGA Gas Analyses

	With K_2CO_3 Catalyst at 800°C	900°C
H_2, volume %	53	29
CO_2, volume %	47	47
CO, volume %	0	22
Total	100	98
Btu/scf	172	169

Table VIII. Proximate Analyses of Ebonite/Catalyst Mixtures

	Ebonite	Ebonite/ 10% $CaCO_3$	Ebonite/ 10% K_2CO_3
Moisture, wt%	2.00	1.70	2.76
Volatile Matter, wt%	37.10	36.53	36.97
Fixed Carbon, wt%	46.43	39.77	37.64
Ash, wt%	14.46	22.00	22.66

Thermogravimetric Analysis Test Results. TGA tests were also conducted on ebonite to determine test matrix conditions for further experiments to be run in the 1- to 4-lb/hr CFBR unit. Ebonite reactivity with steam and two catalysts was investigated at temperatures of 800° and 900°C (Figures 4 and 5). These tests were conducted in a manner similar to those described for automotive shredder residue, with the exception that, in all ebonite TGA tests, the reactant mixtures contained -60-mesh ebonite. A -60-mesh ebonite was used due to the difficulty in obtaining samples of reproducible size distribution from the bulk sample, which was used in CFBR tests. The +60-mesh-size fraction had the same TGA proximate analysis as the -60-mesh fraction, implying the difference (if any) in the reactivities of these two fractions would likely not be chemical, but due to a difference in surface area.

As in the automotive shredder residue TGA tests, to determine the point at which half of the fixed carbon had been converted, proximate analyses were carried out on each sample prior to the reactivity tests. These analyses (Table VIII) showed that volatiles are removed from the reactant mixtures, and the remaining material consists of about 63 percent fixed carbon and 37 percent ash for ebonite with catalyst and about 76 percent fixed carbon and 23 percent ash for ebonite without catalyst (Table VIII). The y-axis on the ASR TGA tests are slightly different than for the ebonite tests. The y-axis on the ASR tests are from zero to 100 weight percent. The ebonite tests were scaled differently and are from 0 to 180 weight percent. During the ebonite tests, when devolatilization and drying were essentially complete, the y-axis was reset to 100 percent. The y-axis on the ASR tests were done in this manner for ease of comparison to another set of test data.

In all of the ebonite tests, the reactant mixtures were heated to the desired reaction temperature and held at temperature until approximately half of the fixed carbon in the sample had been converted to gas, at which point the reaction was terminated by cutting off the steam and heat supply. As stated, each reactivity test was terminated when about half of each devolatilized sample had been converted to gas: 31.5 weight percent of the sample in the case with catalyst, and 38.1 weight percent of the sample in the case with no catalyst. After the termination of heat and steam, the conversion line still continued until the instrument was turned off, since some residual steam was still present, but it was not linear.

The catalysts investigated were Paris limestone (calcium carbonate) and potassium carbonate. All catalyst tests were performed using mixtures of ebonite and 10 weight percent-added catalyst. The ebonite/limestone test was performed at 900°C, and ebonite/potassium carbonate tests were performed at 800° and 900°C. The TGA data indicated that Paris limestone had a minor effect on the reactivity of the ebonite at 900°C (Figure 4, Line E), compared with the reactivity of ebonite without catalyst at the same temperature (Figure 4, Line D). Potassium carbonate, however, significantly affected reactivity. Conversion at 800° (Figure 5, Line D) and at 900°C (Figure 5, Line E) with a potassium carbonate catalyst occurred quite rapidly. The residence time required for complete conversion with this catalyst at 800°C is 5.5 minutes

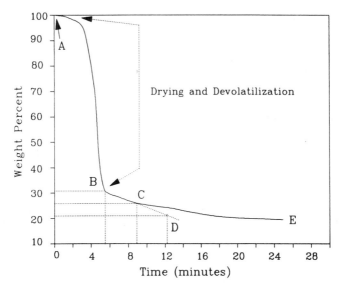

Figure 3. Automotive Shredder Residue at 800°C - With K_2CO_3

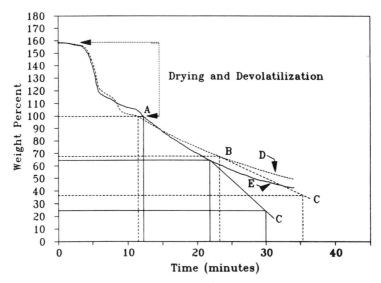

Figure 4. Ebonite at 900°C - With and Without Catalyst

and at 900°C is 2 minutes, whereas conversion time at 900°C without a catalyst is 18 minutes. Residence times for 50 percent and 100 percent conversion were found graphically. The point for 100 percent conversion was found by extrapolating the linear portion of the conversion line. The reaction appears to be zeroth order with respect to carbon. As conversion approaches 100 percent, the reaction is no longer strictly zeroth order because unreactable material (ash) limits access to carbon, but the order goes up only to approximately 0.2, introducing a very small error into the calculated time for total conversion.

Ebonite Continuous Fluid-Bed Reactor Gasification Tests. Bench-scale testing was performed on ebonite in a 1- to 4-lb/hr continuous fluid-bed reactor (CFBR) system, shown in Figure 6. Preheated gas and steam are introduced into the bottom of a 3-inch-diameter reactor. The lower section of the reactor, which is attached to the coal feed system, is made of 3-inch pipe and is 33 inches in length. The freeboard section is made of 4-inch pipe and is 18.75 inches in length. Solids remain in the bed until, through weight loss from gasification, they reach the top of the 3-inch section and fall out through the top bed drain leg, where they are collected in an accumulation vessel. Unreacted fines and some ash particles are entrained and separated from the gas stream by a 3-inch cyclone. Liquids are condensed in one of two parallel, indirect-cooled condensation trains. Gas is then metered and sampled by an on-line mass spectrometer.

Carbon conversion for the ebonite was found to be approximately 90 percent at 900°C, with most of the unreacted ebonite found in the condensation train, indicating that fines blew out of the bed before having sufficient residence time for complete conversion. A narrower particle size for the feed, a lower fluidization velocity, or a larger diameter freeboard section would most likely raise this conversion by reducing fines entrainment. Alternatively, a reactor/cyclone recycle system that is designed for this particular feedstock would also produce higher conversions. Comparing the amount of material in the bed with the feed rate indicates that the residence time for the test was less than one hour. The residence time is extremely dependent on temperature and heatup rate. Ebonite agglomerates at temperatures below approximately 800°C. If the reactor is not above 850°C and at a high heatup rate, the ebonite will agglomerate, greatly reducing the reaction rate and the overall conversion.

Gas produced from gasification and from water-gas shift reactions is between 220 to 280 lbs per 100 lbs of moisture- and ash-free ebonite feed material. The average composition of the product gas is shown in Table IX. Gas produced has a Btu content of approximately 260 Btu/scf. This number does not include nitrogen used in fluidization. Btu content will be lower when inert gas is included, but since the amount of inert gas is process-specific, Btu content of gas produced only is given.

Water conversion was found to be 1.5-2.0 mole water/mole fixed carbon based on material balance data. Trace element analysis showed considerable loss of lead from the ebonite, going from 660 ppm in the feed to 257 ppm in

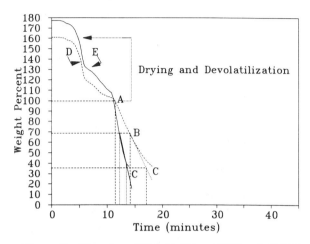

Figure 5. Ebonite - With K_2CO_3 - 800° and 900°C

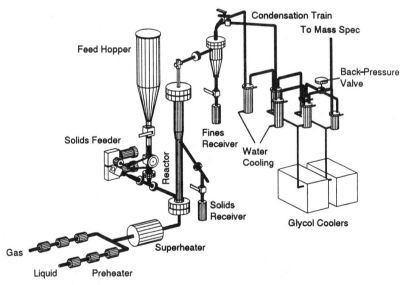

Figure 6. 1- to 4-lb/hr Continuous Fluid-Bed Reactor

Table IX. Ebonite Gas Analysis[a]

H_2	50.1 vol%
CO_2	28.8 vol%
H_2S	0.9 vol%
CH_4	4.5 vol%
CO	15.0 vol%

[a] Without nitrogen fluidizing gas.

the product char. Antimony also decreased considerably, starting off at 696 ppm and ending up at 129 ppm. Chlorine content decreased from 160 ppm to 149 ppm.

Environmental Impacts of Gasifying Waste Material

Gasification of waste materials offers not only the benefit of energy production, but also decreases waste volume that needs to be landfilled. For ebonite, density of the feed material is approximately 0.73 g/mL. Density of the reacted material (top bed drain) is approximately 0.56 g/mL. On an as-received basis, 12.8 m^3 out of 100 m^3 fed will be left over for landfill (assuming the 90 percent conversion). If 100 percent conversion is achieved, 14.5 grams per 100 grams of feed will be left over to landfill, resulting in a volume decrease of 87 percent. ASR density is approximately 0.32 g/cm^3. After reaction, this density is estimated to be about 1.6 g/cm^3, resulting in a volume reduction of about 80 percent.

Wastewater from the process may contain some heavy metals, including lead and antimony. Acid leaching the ebonite prior to gasification may be desirable to eliminate as much of the heavy metals as possible in downstream operations. Additionally, leachability studies will need to be conducted on the unconverted material. Gas cleanup problems will include the need to eliminate sulfur- and chlorine-containing compounds such as H_2S and HCl from the gas stream, as well as trace metals that may vaporize and recondense as particulates. The ebonite feed material is fairly high in sulfur, having an as-received sulfur content of 3.9 percent. The opportunity for greater halide removal in reduced forms may be one of the most important overriding factors for using a gasifier instead of a combustor. For example, HCl is over 50 times more soluble in H_2O than Cl_2 (*3*).

Summary

The future of waste disposal appears to be moving away from landfilling and incineration and toward recycling and using waste materials as a source of energy. New technologies in pollution control and in energy generation techniques, such as gasification, make energy production from waste materials an environmentally acceptable alternative. Tests on ebonite, a hard rubber waste material, and ASR, a waste product of automotive shredders, indicate that a high Btu gas can be produced in short residence times. Considerable work remains to be done to optimize operating conditions and to determine how to deal best with heavy metals, chlorine, and sulfur compounds.

Literature Cited

1. Schmitt, R. J. *Automobile Shredder Residue - The Problem and Potential Solutions*; CMP Report No. 90-1; Center for Materials Production: Pittsburgh, PA, 1990; pp iii, 1-1.
2. Hubble, W. S.; Most, I. G.; Wolman, M. R. *Investigation of the Energy Value of Automobile Shredder Residue*; USDOE Contract No. DE-AC07-84ID12551; EnerGroup, Inc.: Portland, ME, 1987; pp VI-8-8, VI-8-14, VI-8-23.
3. *Perry's Chemical Engineers' Handbook, 6th Edition*; Perry, R. H.; Green, D. W.; Maloney, J. O., Eds.; McGraw-Hill Book Co.: New York, NY, 1973; pp 3-97 - 3-103.

RECEIVED June 22, 1992